T0135716

Low Complexity Physical Layer Cooperation Concepts for Mobile Ad Hoc Networks

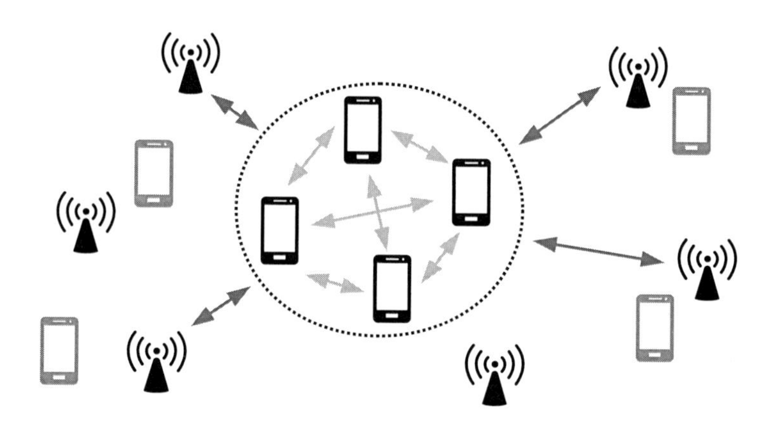

Tim Rüegg

λογος

Series in Wireless Communications
edited by:
Prof. Dr. Armin Wittneben
Eidgenössische Technische Hochschule
Institut für Kommunikationstechnik
Sternwartstr. 7
CH-8092 Zürich

E-Mail: wittneben@nari.ee.ethz.ch
Url: http://www.nari.ee.ethz.ch/

Bibliographic information published by the Deutsche Nationalbibliothek

The Deutsche Nationalbibliothek lists this publication in the
Deutsche Nationalbibliografie; detailed bibliographic data are
available in the Internet at http://dnb.d-nb.de .

ISBN 978-3-8325-4801-8
ISSN 1611-2970

Logos Verlag Berlin GmbH
Comeniushof, Gubener Str. 47,
10243 Berlin
Tel.: +49 030 42 85 10 90
Fax: +49 030 42 85 10 92
INTERNET: http://www.logos-verlag.de

Diss. ETH No. 25366

Low Complexity Physical Layer Cooperation Concepts for Mobile Ad Hoc Networks

A thesis submitted to attain the degree of
DOCTOR OF SCIENCES of ETH ZURICH
(Dr. sc. ETH Zurich)

presented by

TIM RÜEGG

Master of Science (MSc.), ETH Zurich
born on March 16, 1987
citizen of Eschenbach (SG), Switzerland

accepted on the recommendation of
Prof. Dr. Armin Wittneben, examiner
Prof. Dr. Markus Rupp, co-examiner

2018

Day of Doctoral Examination: August 27, 2018

Abstract

Due to the rapid development of wireless communication technologies, its widespread distribution in our daily life, and emerging applications such as ultra high definition mobile video streaming, autonomous driving, or the internet of things with billions of connected devices, future wireless networks face high demands on achievable data rates, coverage, connectivity and delay. Physical layer cooperative communication has a huge potential to meet these demands. By forming a virtual antenna array among multiple nodes, the array gain, the diversity gain and the spatial multiplexing gain of multi-antenna systems can be exploited. This allows to strongly improve the performance of mobile wireless networks.

In this thesis physical layer cooperative communication schemes are investigated to improve the spectral efficiency, the scalability and the coverage range of mobile ad hoc networks (MANETs). We thereby focus on systems with low complexity. Specifically, we investigate multistage cooperative broadcasting to efficiently share common information with a large number of nodes, and provide an accurate performance prediction to it. We furthermore introduce leakage based beam shaping to increase the transmission range, while the signal in undesired directions is suppressed. Moreover, we propose simple resource allocation algorithms for quantize-and-forward receive cooperation to increase the spectral efficiency. We provide theoretical analysis and numerical evaluations of these schemes and investigate their performance, respectively the performance of combinations thereof, in two different scenarios: military MANETs and urban traffic hotspots with ultra high user density. We thereby show that with the proposed schemes the scalability, the coverage range and the throughput of wireless mobile networks can be strongly increased in the considered setups at low complexity and limited delay. Considering, e.g., transmit/receive virtual multiple-input multiple-output (MIMO) for traffic offloading to residential WLAN access points in urban hotspots, very large gains can be achieved compared to non-cooperating schemes if the local exchange is done out-of-band, e.g., in the 60 GHz band. Due to the large available bandwidth, very high exchange rates can be attained. Thereby, cooperative broadcasting is shown to be a key enabler for an efficient data exchange, as it can circumvent the high path loss and the strong shadowing at 60 GHz, even with omnidirectional antennas.

In addition to the cooperative communication schemes, we investigate the relation between the applied transmit power and the resulting interference power at unintended users (so called leakage power) in leakage based precoding, a promising multi-user MIMO precoding approach. Based on these investigations, we propose a target rate precoding as well as a rate optimal precoding, and provide a quasi closed-form solution for both. These precoding schemes lead to promising results and allow to optimize the network performance in terms of outage probability and throughput. They are well applicable in both, MANETs as well as cellular networks with coordinated multipoint transmission.

Kurzfassung

Die rasante technische Entwicklung im Bereich der drahtlosen Kommunikation, ihre weite Verbreitung im Alltag, sowie aufstrebende Technologien und Anwendungen wie ultra-hochauflösendes Video-Streaming, autonomes Fahren und das Internet der Dinge mit Milliarden von verbundenen Geräten, stellen hohe Anforderungen an zukünftige drahtlose Netzwerke im Bezug auf erreichbare Datenraten, Abdeckung, Konnektivität und Latenz. Kooperation auf der Bitübertragunsschicht hat ein enormes Potential diese Anforderungen zu erfüllen. Durch die Bildung von virtuellen Antennenarrays zwischen mehreren mobilen Knoten können die Vorteile, welche Multi-Antennen-System bieten (wie Gruppengewinn, Diversitätsgewinn und räumlicher Multiplexgewinn), gezielt ausgenutzt und so die Leistung von drahtlosen Netzwerken stark erhöht werden.

In dieser Dissertation werden kooperative Kommunikationsverfahren untersucht um die spektrale Effizienz, die Skalierbarkeit sowie die Abdeckung in mobilen ad hoc Netzwerken (MANETs) zu erhöhen. Wir fokussieren uns dabei auf Systeme mit geringer Komplexität. Konkret konzentrieren wir uns auf *Multistage Cooperative Broadcast*, womit eine Nachricht mit einer grosse Anzahl von Knoten effizient geteilt werden kann, und präsentieren eine präzise Ausbreitungs-Prognose dafür. Weiter führen wir *Leakage Based Beam Shaping* ein um die Kommunikationsreichweite zu erhöhen, während das Signal in ungewollte Richtungen unterdrückt wird. Zudem schlagen wir simple Ressourcenzuordnungsalgorithmen für die *Quantize-and-Forward* Empfangskooperation zur Erhöhung der spektralen Effizienz vor. Wir präsentieren theoretische wie auch numerische Auswertungen dieser Verfahren und untersuchen ihr Verhalten, respektive das Verhalten von Kombinationen dieser Verfahren, in zwei verschiedenen Szenarien: militärische MANETs und städtische Hotspots mit hoher Knotendichte. Dabei zeigen wir, dass mit den vorgeschlagenen Verfahren die Skalierbarkeit, die Abdeckung und der Durchsatz von mobilen drahtlosen Netzwerken in den erwähnten Szenarien bei kleiner Latenz und limitierter Komplexität stark erhöht werden kann. In städtischen Hotspots zum Beispiel können mit virtuellen Sende-/Empfangs-Antennenarrays sehr grosse Gewinne erzielt werden, wenn der Datenaustausch zwischen den kooperierenden Knoten in einem separaten Frequenzband, z.B. im 60 GHz Band, erfolgt. Durch hohe verfügbare Bandbreiten können sehr hohe Austauschraten erzielt werden, wobei *Cooperative Broadcast* ein Schlüsselelement für den effizienten Datenaustausch ist. Trotz omnidirektionaler Antennen können damit sowohl die hohe Ausbreitungsdämpfung (*Path Loss*) als auch die starke Abschattung (*Shadowing*) bei 60 GHz umgangen werden.

Des Weiteren untersuchen wir für *Leakage Based Precoding*, eine vielversprechenden Mehrfachnutzer-Vorcodierung, den Zusammenhang zwischen aufgebrachter Sendeleistung und generierter Interferenzleistung bei unbeteiligten Knoten (sogenannte *Leakage*-Leistung). Basierend auf diesen Untersuchungen schlagen wir sowohl eine Zielraten-Vorcodierung als auch eine ratenoptimale Vorcodierung vor und präsentieren eine quasi-analytische Lösung dafür. Diese Vorcodierungsverfahren erzielen vielversprechende Resultate und erlauben es, die Netzwerkperformance im Bezug auf Ausfallwahrscheinlichkeit und Durchsatz zu optimieren. Sie sind sowohl in MANETs als auch in zellulären Netzwerken mit *Coordinated Multipoint Transmission* anwendbar.

Contents

Abstract i

Kurzfassung iii

1. Introduction 1
   1.1. Cooperative Communication in Mobile Wireless Networks 2
   1.2. Infrastructure-less Wireless Networks 5
   1.3. Infrastructure-based Wireless Networks 6
   1.4. Contributions and Outline 7

2. Cooperative Broadcasting in Mobile Ad Hoc Networks 11
   2.1. Classical Broadcasting Methods 11
   2.2. Multistage Cooperative Broadcast 12
   2.3. Scalability of Routing Overhead 14
   2.4. Conclusion 16

3. Stochastic Geometry Based Performance Prediction 17
   3.1. Related Work 18
   3.2. System Model 19
   3.3. Inter-Node Distances 20
   3.4. Performance Prediction of Multistage Cooperative Broadcast 25
   3.5. Multi Hop Transmission Performance Prediction 34
   3.6. Conclusion 39

4. Leakage Based Multi-User MIMO Precoding 41
   4.1. Related Work 43
   4.2. System Model 44
   4.3. Rate Optimal Leakage Based Precoding 45
   4.4. Leakage Based Target Rate Precoding 56
   4.5. Conclusions 66

5. Leakage Based Beam Shaping 67
5.1. Related Work . . . 68
5.2. Scenario and System Setup . . . 69
5.3. Pattern Synthesis . . . 71
5.4. Performance Evaluation . . . 78
5.5. Practical Considerations and Discussion . . . 82
5.6. Conclusions . . . 87

6. Quantize-and-Forward Virtual MIMO Receive Cooperation 89
6.1. Related Work . . . 90
6.2. System Model and Problem Formulation . . . 91
6.3. Backhaul Resource Allocation . . . 93
6.4. Relay Selection . . . 95
6.5. Cascade Resource Allocation . . . 102
6.6. Conclusions . . . 107

7. Cooperative Communication in Military Mobile Ad Hoc Networks 109
7.1. Range Extension . . . 110
7.2. Spatial Multiplexing . . . 115
7.3. Conclusions . . . 119

8. User Cooperation Enabled Traffic Offloading in Urban Hotspots 121
8.1. Related Work . . . 122
8.2. System Setup . . . 123
8.3. User Cooperation for Ultra Dense Environments . . . 124
8.4. Simulation Framework . . . 127
8.5. Uplink Performance Evaluation . . . 128
8.6. Downlink Performance Evaluation . . . 133
8.7. Spatial Reuse and Local User Performance . . . 139
8.8. Conclusions . . . 145

9. Summary and Conclusion 147

A. Appendix 149
A.1. Directive Antenna Pattern . . . 149
A.2. WINNER II, Scenario C2 Channel Model . . . 149
A.3. 60 GHz Log-Distance Path Loss Model . . . 150

A.4. Monotony of Achievable Rates of Leakage Based Precoding 151
A.5. Derivation of the Partial Derivative of the Cascade Decoding Rate . . . 153

List of Figures **155**

Acronyms **159**

Notation and Variables **161**

Bibliography **167**

Curriculum Vitae **177**

1 Introduction

Ever since the first experiments with a radio-telegraph system by Guglielmo Marconi in the 1890ties, the technology of wireless communication via radio waves experienced a rapid development. From wireless voice transmission over television broadcasting to cellular networks and satellite based communication, various applications emerged. Due to the introduction of digital communication, increased computational power and the miniaturization of the hardware in the last decades, wireless communication became omnipresent. Nowadays, devices and technology such as smart phones, satellite based localization or body area networks (bluetooth head phones, smart watches, fitness trackers) determine our daily life and became an important business factor. However, the rapid development of the technology, its widespread distribution and services such as video telephony, mobile internet and other multi-media applications have led to an exponential growth of mobile data traffic and a strongly increasing number of connected devices in the last years [1]. Emerging technologies like autonomous driving with vehicle to vehicle communication [2] and the internet of things [3] with billions of connected devices even further increase the demands on future wireless networks. Data rates and spectral efficiencies which are larger by an order of magnitude than in nowadays systems have to be achieved, coverage increased, delay reduced and massive connectivity provided [1].

Nowadays wireless networks can be distinguished into infrastructure-less wireless networks, such as mobile ad hoc networks (MANETs) or wireless sensors networks, and infrastructure-based wireless networks, such as cellular networks or satellite networks. Depending on the application and the environment, a wireless network is either implemented as a infrastructure-less or infrastructure-based wireless network. However, for both architectures new technologies and approaches have to be introduced to face the tremendous demands on future wireless networks.

Among others, physical layer cooperation between wireless nodes has a huge potential. By forming a virtual antenna array with distributed signal processing, the gains offered by multi-antenna techniques can be exploited. Therewith, the efficiency, the throughput and the scalability of both, infrastructure-less and infrastructure-based wireless networks can be strongly increased.

To this end, we investigate physical layer cooperation schemes in this thesis and apply it to both types of networks. In the remainder of this chapter we provide an overview on multi-antenna techniques and their application in cooperative communication in Sec. 1.1, introduce infrastructure-less wireless networks with their promises and challenges in Sec. 1.2 and discuss the same for infrastructure-based wireless networks in Sec. 1.3. In Sec. 1.4 finally, we summarize the contributions and provide an outline of this thesis.

1.1 Cooperative Communication in Mobile Wireless Networks

In cooperative communication individual users support each other in order to improve the individual and/or the network performance. In this thesis, we focus on the approach where multiple nodes form a virtual antenna array and then jointly transmit or receive. This way, they can exploit all benefits of multi-antenna systems, potentially leading to better coverage, higher throughput, higher connectivity and better scalability.

1.1.1 Multi-Antenna Systems

Multi-antenna techniques have been shown to have a huge potential in wireless communications by the landmark papers of Foschini [4], Telatar [5], Alamouti [6], and Tarokh et al. [7] in the late 1990ties. Besides the temporal and spectral domain, multi-antenna systems can also exploit the spatial domain, and therewith achieve higher signal power at the receiver, lower error rates and higher bit rates [8], also denoted as *array gain*, *diversity gain* and *spatial multiplexing gain*. A broad overview on multi-antenna techniques for wireless communications is given in [8], along which we summarize the potential and benefits in the following.

Beam shaping Similar to the concepts of phased arrays [9], the radiation pattern of the antenna array can be designed such that the signal energy is concentrated in a desired direction while suppressed in undesired directions. That is, the signal-to-noise ratio (SNR) at a receiver can be increased while interference to other nodes can be mitigated. To this end, the same signal is transmitted from multiple antennas, just individually phase-shifted and weighted (in contrast to classical phased arrays where no weighting is done), such that the desired radiation pattern is obtained. The same principles can also be applied on the receiver side by emphasizing the received signal from a desired direction while suppressing other directions. A comprehensive overview on pattern design methods and their application in wireless mobile communications can be found in [10,11]. To design the radiation patterns no channel state information is necessary, only the relative location information of the transmit respectively receive antennas.

Diversity By transmitting multiple versions of the desired signal distributed over multiple antennas, a diversity gain can be achieved at the receiver. Different to single antenna systems, the redundancy can not only be introduced in the temporal and spectral domain but also in the spatial domain by so called space-time codes. With such codes a diversity gain can be achieved without lowering the effective bit rate. Since the initial papers of Alamouti [6] and Tarokh et al. [7], a large variety of space-time-codes has been proposed [12]. To achieve diversity on the receiver side well established receive diversity techniques based on linear combining can be applied [13].

Spatial multiplexing Compared to single antenna nodes, the bit rate between a transmitter and a receiver, each with M sufficiently spaced antennas, can be increased by a factor M without additional transmit power or bandwidth if the signal arrives with sufficiently different spatial signature at the receive antennas (i.e. in a rich scattering environment and/or with sufficient angular spread between transmit and receive antennas). This gain in rate is called spatial multiplexing gain [4,5,14]. With a transmit array of M antennas it is also possible to send M orthogonal streams and thus serve up to M independent users simultaneously. To this end, appropriate precoding, so called multi-user multiple-input multiple-output (MIMO) precoding, is necessary [15, 16]. These precoding schemes suppress the co-channel interference among the independent nodes such that each node can decode its signal in the presence of only limited or no interference. Prominent representatives thereof are zero-forcing [17], which completely nulls the co-channel interference, and SLNR precoding [18], which maximizes the signal-to-leakage-plus-noise ratio (SLNR) for each node. Similarly, a receive array with M antennas can decode up to M streams of independent users simultaneously. To this end, joint decoding can be applied.

Different multi-antenna techniques can also be combined, e.g., spatial multiplexing with diversity. However, there is a fundamental trade-off between the achievable spatial multiplexing gain and the achievable diversity gain [19].

1.1.2 Cooperative Communication

All the potential gains of multi-antenna systems as discussed above can also be achieved by transmit or receive cooperation of a cluster of single antenna nodes. To this end, they form a virtual antenna array and jointly transmit or receive. That is, they either have to exchange their transmit data first in order to apply a transmit cooperation scheme, or they share their received signals with each other in order to process the data in receive cooperation. Cooperative communication aiming at the benefits of multi-antenna systems it is often also called virtual MIMO or cooperative MIMO.

The large potential gains of cooperative communication has drawn extensive interest in research and many schemes and approaches have been proposed, e.g., [20–23]. In [21] for example, a cooperative broadcasting protocol exploiting the diversity gain has been presented, which allows fast spreading of common information among a large number of nodes. Exploiting the spatial multiplexing gain, e.g., mobile stations in a cellular network are proposed to form a virtual antenna array for capacity enhancement in [22]. They are then jointly accessed by the base stations and exchange their received data among each other, in order to decode the messages. In [23], it is shown that hierarchical user cooperation can achieve linear capacity scaling in ad hoc networks. That is, the total throughput in the network increases linearly with increasing number of nodes. To this end, the involved nodes hierarchically apply a 3 phase protocol. In phase one, the nodes in a source cluster locally exchange their transmit data. In phase two, these nodes then jointly transmit this data to the destination cluster, where in phase 3, the received signals are again locally exchanged among the involved nodes in the cluster and finally decoded.

All these approaches promise high gains in network capacity, throughput or scalability. However, the performance of virtual MIMO schemes strongly depends on the efficiency of the data exchange among the cooperating nodes. In [24], e.g., it is shown under which conditions cooperative MIMO increases the channel capacity between two clusters of cooperating nodes. Furthermore, the coordination of the cooperation, the medium access control and the synchronization of the nodes are crucial aspects of cooperative communication.

1.2 Infrastructure-less Wireless Networks

As the name indicates, infrastructure-less wireless networks do not rely on a fixed infrastructure. Instead, the nodes communicate directly with each other without a central administration such as an access point which coordinates the network traffic. The pairwise communication between two nodes is done either with direct communication or with a multi hop scheme if the destination cannot be reached within a single hop. Thereby, all nodes have all functionalities and are hosts as well as routers at the same time [25]. Due to this flexibility, infrastructure-less wireless networks have the potential to adapt fast to new circumstances as, e.g., the outage of nodes and the circumvention of bottlenecks. Furthermore, as no external infrastructure is required, they can be set up and provide local coverage instantly.

Due to the improvised nature of infrastructure-less wireless networks, they are also called *wireless ad hoc networks*. Two major types of wireless ad hoc networks can be distinguished: mobile ad hoc networks (MANETs) and wireless sensor networks [26]. A MANET is a self-configuring wireless network which consists of multiple randomly moving mobile nodes. Typical application scenarios of MANETs are vehicular to vehicular communication, emergency situations, post-disaster communication or military networks. In wireless sensor networks, the nodes typically have lower or no mobility and much lower resources available (energy, memory, computational power, bandwidth). However, both network types suffer from the same drawbacks. To overcome large distances multi hop communication is performed. As no central unit coordinates the network, sophisticated distributed routing protocols and medium access protocols are necessary, as well as distributed synchronization among the nodes. This decentralized coordination leads to poor scalability for increasing number of nodes in the network. The reason for this is the strongly increasing overhead of the message routing as well as the medium access control [25–28]. Furthermore, multi hop communication introduces long delays and is prone to node outages and bottlenecks.

Cooperative communication as discussed in Sec. 1.1.2 can improve the performance of ad hoc networks in various ways. With a cooperative broadcast protocol [21] the routing overhead in ad hoc networks can be decreased by up to an order of magnitude [29]. Thus the scalability with the number of nodes is increased. Forming a virtual antenna array and optimizing the radiation pattern, larger distances can be overcome with a single hop. Hence, compared to multi hop communication, the delay can be reduced, the routing overhead decreased, and a higher robustness against bottlenecks and node failures can be provided. With transmit and receive virtual MIMO, multiple streams can be transmitted simultaneously and the spectral efficiency can be increased [24]. Furthermore, a better scalability can be achieved as shown in [23].

1.3 Infrastructure-based Wireless Networks

In contrast to infrastructure-less wireless networks, infrastructure-based wireless networks rely on a fixed infrastructure. Thereby, the mobile wireless nodes connect to fixed access points, via which they can communicate with each other or access the internet. That is, all traffic is routed via the access points. Typical examples of infrastructure-based wireless networks are wireless local-area networks (WLANs), cellular networks or satellite networks.

To face the tremendous demands on wireless networks (especially in cellular networks) [1], various candidate technologies have been presented for the next generation networks, such as massive MIMO systems [30], millimeter wave (mmW) communication [31], or network densification with heterogeneous networks and traffic offloading to WLANs [32]. Large gains can also be achieved by cooperative communication such as coordinated multipoint (CoMP) transmission or cloud radio access networks (CRAN), where the cooperation is done among the infrastructure nodes to serve more mobile nodes simultaneously with higher quality of service.

Due to the centralized coordination and the fixed network structure, infrastructure-based networks have a superior scalability. For an increasing number of mobile nodes simply more access points can be installed, such that the network capacity scales linearly with the number of nodes. Nevertheless, even with the approaches discussed above, serving the users in a traffic hotspot, i.e. an area with ultra high user density such as a busy public square or a train station, still remains a problem. While in theory the capacity scales linearly with the number of transmit and receive antennas, cellular networks become capacity limited in dense environments as a lot of users are assigned to one or a few base stations. Further base station sites would have to be identified to increase the throughput, which frequently turns out to be impossible.

In such a scenario user cooperation protocols as discussed for ad hoc networks could improve the performance. By exploiting the spatial multiplexing gain and the array gain of virtual antenna arrays formed by the users in the hotspot the traffic can be distributed over a wider area.

1.4 Contributions and Outline

By exploiting the array gain, the diversity gain, and the spatial multiplexing gain which multi-antenna systems provide, cooperative communication has the potential to strongly improve the performance of ad hoc networks by forming virtual antenna arrays of multiple nodes. The same schemes can be applied to serve users in a traffic hotspot in cellular networks. To this end, we investigate low complexity physical layer cooperative communication schemes to improve the spectral efficiency, the scalability and the coverage range of ad hoc networks in this thesis. Specifically we focus on the following three schemes:

- Cooperative broadcasting has been shown to be an efficient scheme to share common information with a large number of nodes outside of the one-hop communication range by exploiting the diversity gain [21]. Hence, it has the potential to strongly reduce the routing overhead in ad hoc networks. The performance prediction of this scheme for large networks can be very expensive computation-wise. However, for both network design and network operation (reconfigurability) efficient approximations of the performance are of paramount importance. To this end, we propose an extension of results of stochastic geometry for this purpose by considering the statistics of the inter-node distances in random networks [33]. At low computational complexity it allows very accurate performance prediction, even at low node density. At the same time, the derived distance distributions are also well suited to predict the performance of multi hop transmission.
- Considering the random node arrangement, we introduce a novel scheme for distributed beam shaping in MANETs [34] exploiting the array gain. It minimizes the leakage power (signal power in undesired directions) while maintaining coherent addition in the desired direction. That is, the transmission range can be extended, while suppressing the interference into the network and reducing the risk of interception in military scenarios.
- In the context of receive cooperation exploiting the spatial multiplexing gain, we propose efficient low complexity resource allocation schemes for quantize-and-forward receive cooperation that optimize the resources used by each cooperating node in the local exchange phase [35].

We provide theoretical analysis and numerical evaluations of these schemes and investigate their performance, respectively the performance of combinations of the schemes, in two different scenarios: military MANETs and urban traffic hotspots with ultra high user density [36–38]. Furthermore, we investigate the relation between transmit power and leakage power in leakage based precoding, a promising multi-user MIMO precoding scheme as a basis for transmit virtual MIMO. Considering this relation, we propose the rate optimal precoding under joint leakage and transmit power constraints and derive an iterative closed-form solution to it [39]. As multi-user MIMO precoding often leads to very unevenly distributed rates among the served nodes, we additionally propose a low complexity but efficient target rate precoding based on the transmit power - leakage power trade-off [40]. The proposed schemes, their performance and the gained insights are presented as follows in the remainder of this thesis.

In Chap. 2, we review cooperative broadcasting, a transmit cooperation scheme to efficiently share common information with a large number of nodes in a MANET. Compared to classical broadcasting schemes large gains in the number of necessary retransmissions can be achieved, and therewith, by reducing the routing overhead, the scalability of MANETs can be strongly improved.

In Chap. 3, we are going to present a novel low complexity but accurate performance prediction of multistage cooperative broadcast and multi hop transmission based on inter-node distance distributions.

In Chap. 4, we investigate the relation between transmit power and leakage power in leakage based precoding and introduce rate optimal leakage based precoding as well as leakage based target rate precoding. The proposed schemes are evaluated in a cellular setup with multi-antenna base stations.

In Chap. 5, leakage based beam shaping, a low complexity pattern synthesis approach based on the maximization of the signal-to-leakage ratio, is proposed and evaluated. Furthermore, various practical considerations of leakage based precoding in a military MANET are addressed.

In Chap. 6, the resource allocation problem of quantize-and-forward receive cooperation is addressed and low complexity resource allocation approaches are presented.

In Chap. 7, we consider the application of cooperative communication schemes in a military MANET, namely multistage cooperative broadcast, beam shaping and transmit virtual MIMO. Their performance is evaluated on a conceptual basis and compared to conventional schemes in terms of transmission range, coverage area and throughput.

In Chap. 8, we combine transmit virtual MIMO with leakage based precoding, quantize-and-forward virtual MIMO receive cooperation and cooperative broadcasting to serve the mobile stations in an urban traffic hotspot, i.e. an area with ultra high user density. The proposed scheme is evaluated in a rigorous feasibility study with realistic parameters, and suitable operating regimes are identified.

In Chap. 9, we finally conclude the thesis.

2

Cooperative Broadcasting in Mobile Ad Hoc Networks

In this chapter, we are going to discuss cooperative broadcasting methods for wireless mobile networks.

Network wide broadcasting, i.e. transmitting a message to all nodes in the network, is a very important functionality in mobile ad hoc networks (MANETs). Many widely employed routing protocols rely on it, such as the destination-sequenced distance-vector (DSDV) protocol, dynamic source routing (DSR) and the zone routing protocol (ZRP) [41]. Hence, to minimize the overhead of setting up and maintaining the routing protocol, an efficient broadcasting scheme is key. In classical broadcast protocols the number of retransmissions of a broadcast message scales poorly with the number of nodes in the network. A promising alternative to these established protocols is a cooperative broadcast, where the nodes which received a message jointly retransmit it. This way, the scalability can be strongly improved and the reliability increased. In the remainder of this chapter we first review classical broadcasting methods in Sec. 2.1 and then introduce cooperative broadcasting in Sec. 2.2. In Sec. 2.3 we discuss the impact of the broadcasting method on the overhead of common routing protocols and Sec. 2.4 finally concludes the chapter.

2.1 Classical Broadcasting Methods

Different methods to broadcast common information to all nodes in the network have been proposed in literature [42]. The most straightforward form is *simple flooding*, where each node retransmits a broadcast message exactly once. That is, the necessary time to broadcast a message scales linearly with the number of nodes. This guarantees a high reliability of the broadcast message to reach all nodes in the network (if reachable at all). However, it is very inefficient, as probably many of the retransmissions could be mitigated without decreasing the coverage.

Various proposals have been made to improve the efficiency of simple flooding, as summarized, e.g., in [42]. In *probability-based methods* the nodes only forward the message with a certain probability depending on the network topology. In *area-based methods* a node only retransmits a broadcast message if the newly covered area is sufficiently high. In *neighborhood-based methods* the decision whether to broadcast or not is based on the comparison of the list of the neighbors of the source which is added to each message, and the list of the own neighbors. To this end, a list of neighbors has to be kept updated continuously.

With all these methods less nodes are retransmitting and thus less time is necessary for the broadcast. An evaluation of representatives of these schemes is provided in [42] for a constant network area and various node densities. While the number of retransmissions is decreased, the probability- and area-based methods also decrease the reliability of the protocol to reach all nodes in the network. Furthermore, the necessary time to broadcast a message still scales linearly with the number of nodes if the network is growing in the network area with constant node density (the number of retransmissions increases linearly with the area for any scheme and thus linearly in the number of nodes).

Besides their inefficiency in broadcasting a message, broadcasting protocols also jitter the retransmission of the broadcast message in order to prevent collisions [42]. That is, they delay the retransmission by a random amount of time, such that, e.g., a carrier sense multiple access/collision avoidance (CSMA/CA) protocol can be applied. However, the hidden node problem is still present. Hence, the inefficiency in broadcasting a message is even further increased by the medium access control necessary to execute the broadcast protocol and the collisions which still occur due to the hidden node problem.

2.2 Multistage Cooperative Broadcast

In order to increase the efficiency of broadcasting a message to all nodes in the network, a cooperative scheme can be applied such as dynamic cooperative broadcasting or multistage cooperative broadcasting.

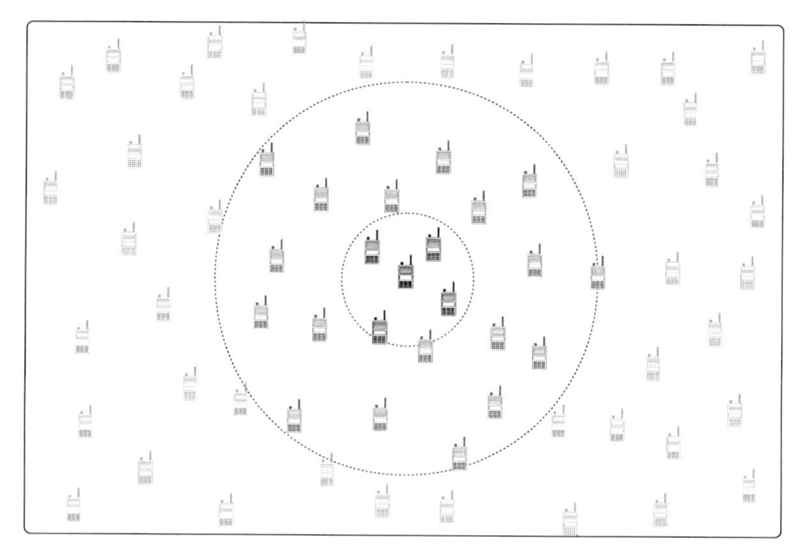

Figure 2.1.: Illustration of the multistage cooperative broadcast: In a first step, the source (blue) transmits its message. All nodes which could decode (purple) jointly transmit with the source in the second time slot, reaching the red nodes, and so on until all nodes could decode the message.

Dynamic cooperative broadcast In the dynamic cooperative broadcasting as introduced in [43] (also called *dynamic decode and forward collaborative communications*), the initializing node starts to transmit its message while all other nodes are in receiving mode. As soon as a receiving node can decode the message it becomes a relay. To this end, it re-encodes the message using a different codebook and supports the initializing node by transmitting its codeword as well. This is done until all nodes could decode the message. Hence, the message spreads out like a wave from the initializing node to all receiving nodes. As all nodes use a different codebook, the signal contributions add up in power at the receiving nodes. That is, for each node becoming a relay, the instantaneous achievable rate at the receiving nodes is increased. This allows to efficiently share a message with all other nodes in the network. However, the complexity is very high. Each node would have to know the codebook of all transmitting nodes as well as the time they start to transmit.

Multistage cooperative broadcast A more practical form is the so called multistage cooperative broadcast with fixed time slots and fixed transmission rates [21]. In a first time slot (corresponding to the first stage respectively the first hop of the protocol) the source transmits its message. All nodes which have a sufficient signal-to-noise-ratio (SNR) decode the message and retransmit it in the next time slot (i.e. in the next stage respectively hop). This procedure is repeated until all intended nodes could decode the message, and is illustrated in Fig. 2.1. The number of retransmissions of the message by each node which could decode can be set arbitrarily (e.g., each node retransmits it exactly once, or each node retransmits it in all subsequent time slots until the broadcast terminates). The retransmission is thereby performed using a transmit diversity scheme such that the individual signal contributions add up in power at the receiving nodes. Hence, large SNRs can be achieved at the receivers and the message spreads quickly through the network. To achieve the diversity gain at the receiving nodes any transmit diversity scheme such as different codebooks for all transmitters, applying phase rolling [44], or a space-time code would be sufficient. An overview over such space-time codes can be found in [12].

Compared to classical broadcasting large gains can be achieved in throughput as well as in delay. A performance evaluation illustrating the fast spreading of multistage cooperative broadcast can be found in Chap. 3 and Chap. 7. As a node which received the message simply retransmits it in the next time slot, no routing or special medium access control is necessary. Hence, the loss in efficiency due to the jitter in non-cooperating broadcasting schemes is also mitigated.

Multistage cooperative broadcasting is not only well suited to efficiently broadcast a message to a large number of nodes, but could also be used for the communication between two distant nodes as an alternative to the multi hop communication. However, the price of shorter delay and no necessity for a routing protocol is the increased energy consumption, as many nodes transmit over multiple time slots. Furthermore, the whole network is blocked by one transmission.

2.3 Scalability of Routing Overhead

Routing protocols can be separated into proactive routing protocols, reactive routing protocols and hybrid protocols [41]. In proactive routing each node has a full routing table available at any time. To obtain these tables, each node periodically broadcasts the list of its neighbors to all other nodes in the network. In reactive routing, the route from the source to the destination is discovered on demand by broadcasting a route request message. Hybrid routing protocols combine proactive and reactive approaches, e.g., by splitting up the network into clusters. Within each cluster a proactive routing is applied, between different clusters a reactive routing.

To visualize the impact of the broadcasting efficiency on the routing overhead, we are going to model the minimal amount of information to exchange for proactive routing in a network with N nodes according to [29,45]. The main contribution to the routing overhead are the so called *topology control messages*. These messages contain the list of neighbors of a node and are exchanged periodically in order to establish and maintain a complete routing table at each node. The minimal overhead of transmitting all topology control messages once can be stated as

$$I = \eta \xi N^2 \log_2 N, \tag{2.1}$$

where η denotes the average number of transmissions necessary to reach all nodes in the network (depending on the broadcasting method) and ξ represents the ratio of the coverage area of one node and the total network area (independent of the broadcasting method). That is, ξN is the average number of neighbors per node. As N nodes broadcast their list of neighbors and as $\log_2 N$ bit are necessary to uniquely identify each node, we end up with I bits to be shared among all nodes. The other contributions to the routing overhead, such as the *hello messages* to obtain the list of neighbors of each node and the messages to be exchanged due to route invalidities, have a smaller impact on the total overhead. They are thus omitted in this simple analysis. Disregarding η, it can be seen that the routing overhead scales with $\mathcal{O}(N^2 \log_2 N)$. Depending on the broadcasting method the overhead scales even worse. For simple flooding $\eta = N$ (assuming each node retransmits the message once in an orthogonal time slot and no collisions occur). Probability based methods can decrease the number of retransmissions, but η still scales linearly with the number of nodes. While neighborhood-based methods and area-based methods can limit η if the network is only growing in N but not in area (i.e. the node density is increasing), η still scales linearly for increasing network area at constant node density. Cooperative broadcast, however, can strongly decrease η. As the nodes jointly retransmit, only a few retransmissions are necessary to cover even very large networks (as shown in the performance evaluation of multistage cooperative broadcast in Chap. 3). For constant network area but increasing number of nodes, η can even decrease. Hence, while the routing overhead still grows significantly with increasing number of nodes, it can be strongly reduced with cooperative broadcast compared to classical broadcasting methods.

An evaluation of the routing overhead comparing simple flooding and multistage cooperative broadcast in military MANETs can be found in [29]. It is shown that the relative routing overhead (i.e. the relative amount of time necessary to establish and maintain the routing protocol) can be reduced by an order of magnitude using multistage cooperative broadcasting instead of simple flooding. This leads to better scalability and allows for larger networks.

2.4 Conclusion

For a fast delivery of a message to all nodes in a MANET, as, e.g., required to establish the routing protocol, an efficient broadcasting is very important. To this end, we introduced multistage cooperative broadcast in this chapter. Compared to conventional broadcasting schemes, large gains can be achieved in terms of necessary number of retransmissions. Furthermore, as the retransmission is carried out by all retransmitting nodes simultaneously, no collision avoidance is necessary, further increasing the efficiency compared to conventional schemes. Using a cooperative broadcast, the scalability of MANETs can be strongly improved, as the relative routing overhead can be decreased by up to an order of magnitude.

3

Stochastic Geometry Based Performance Prediction

In this chapter, we are going to introduce a low complexity performance prediction of multistage cooperative broadcast.

Cooperative broadcasting has been shown to be an efficient scheme to share common information with a large number of nodes outside of the one-hop communication range [21], as, e.g., required to setup and maintain the routing protocol in mobile ad hoc networks (MANETs). Compared to a conventional broadcasting scheme the relative overhead of a proactive routing protocol can be reduced up to an order of magnitude [29] as discussed in Chap. 2. Hence, cooperative broadcasting is a key technology to improve the scalability of MANETs.

The performance prediction of this scheme with Monte Carlo simulations can be very expensive computation-wise for large networks. However, for network design as well as network operation (reconfigurability) efficient approximations of the performance are of paramount importance. In this chapter, we are going to present a novel performance prediction of multistage cooperative broadcast. It is based on inter-node distance distributions and allows for an efficient and accurate performance prediction for large networks. We derive the necessary distance distributions and discuss simple but accurate approximations of their expected values (which can not be found analytically). Using these distances, we then propose a performance prediction for multistage cooperative broadcast which considers each hop separately and employs the expected values of the coverage range (i.e. the range in which the message could be decoded) and the number of nodes reached per hop. The performance of the proposed approach is evaluated for various node densities and compared to Monte Carlo simulations. It already works very accurately for low node densities and has low computational complexity. Furthermore, the derived distance distributions are also well suited to predict the performance of multi hop transmission.

This chapter is mainly based on our work published in [33] and expanded with the performance prediction of multi hop transmission. The remainder is structured as follows. In Sec. 3.1 related work is reviewed and in Sec. 3.2 we introduce the system model. In Sec. 3.3 the necessary distance distributions are derived and approximations of their expected values are presented. In Sec. 3.4 the proposed performance prediction of multistage cooperative broadcast is introduced and evaluated. Analogously, in Sec. 3.5 the performance prediction of multi hop transmission is introduced and evaluated. Sec. 3.6 finally concludes the chapter.

3.1 Related Work

Multistage cooperative broadcasting has been widely studied in literature. In [46], its performance has been investigated in line networks by deriving an analytical expression for the outage probability. In a similar setup, the authors of [47] determine an upper bound on the maximum achievable gain compared to classical broadcasting. In [48], a two-hop cooperative broadcast protocol is presented and the outage probability and the diversity-multiplexing trade-off are developed. In [21], the transmission dynamics of the multistage cooperative broadcast protocol are derived recursively in a two dimensional network for the continuum limit, i.e. by letting the number of nodes grow to infinity while keeping the total transmit power constant. Nevertheless, the performance prediction of multistage cooperative broadcast is still a problem. While the evaluation with numerical simulations is computationally very costly, [46] and [47] only consider line networks and [48] a two-hop network. Although [21] allows to predict the performance very accurately for high node densities, it is not well suited for low densities.

Various papers have recently considered distance distributions based on stochastic geometry in MANETs. An introduction to the mathematical tools of stochastic geometry and random geometric graphs is given in [49]. In [50], a general method is presented to derive the distribution of the distance between the n-th closest node and an arbitrary node of interest for a general binomial point process (BPP). Therewith, the probability density functions (PDFs) of the distance from a source located in the center of a regular polygon to N other nodes uniformly distributed in the polygon are derived. Based on these distributions wireless network characteristics are studied. In [51], the authors derive the complementary cumulative distribution function of

Euclidean distances between randomly selected nodes uniformly distributed on a disk and therewith present tight lower bounds on the outage performance if these nodes communicate with each other. In [52], a stochastic geometric approach is used to characterize the signal-to-interference ratio distribution of a millimeter wave ad hoc network with directional antennas. There are many more papers also considering stochastic geometry to approximate the performance of wireless networks. However, none of the derived distance distributions is suitable for the performance prediction of cooperative broadcast.

3.2 System Model

We consider an ad hoc network where the locations of the nodes are points of a homogeneous Poisson point process (PPP) Φ with intensity δ in $\mathbb{R}^2$. That is, for any subset of the plane of area Ω, the mean of the number of nodes in this subset is given by $\delta \cdot \Omega$. The channel between any two nodes is determined by a distance dependent path loss with path loss coefficient γ and a random phase shift. That is, no small scale fading is considered in order to simplify the performance prediction. The impact of this simplification on the validity of the results is discussed in Sec. 3.4.4 for multistage cooperative broadcast and in Sec. 3.5.4 for multi hop transmission. Considering Gaussian transmit symbols $s \sim \mathcal{CN}(0, P_{\text{Tx}})$ with the transmit power P_{Tx} and additive white Gaussian noise $w \sim \mathcal{CN}(0, \sigma_w^2)$, the received signal of a node at distance d from the source can be written as

$$y = d^{-\gamma/2} \cdot h \cdot s + w, \tag{3.1}$$

with $h = e^{j\theta}$, $\theta \sim \mathcal{U}(0, 2\pi)$. Without loss of generality the source is considered to be located at the origin.

We consider time slots of fixed length and a fixed transmission rate per hop of R_{min} bps/Hz. Whenever a node can decode the message (i.e. when it achieves a decoding rate of at least R_{min}), it supports the source in all consecutive time slots by retransmitting the message using a different codebook, until the message has reached all intended nodes. Due to the different codebooks, the signal contributions add up in power at the receiving nodes, leading to a fast spreading of the message.

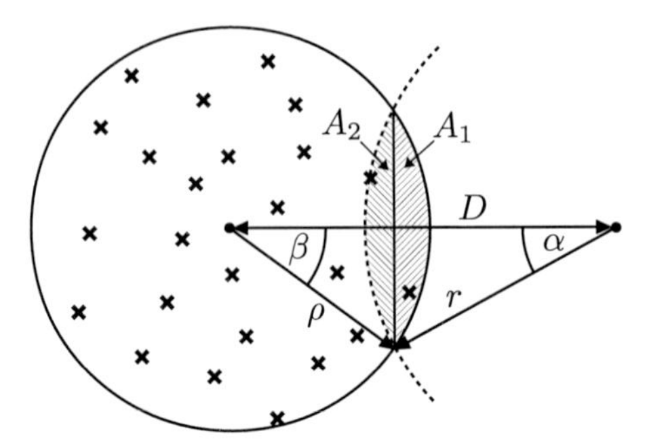

Figure 3.1.: A typical setup for one hop of the cooperative broadcast: N nodes distributed on a disk of radius ρ jointly transmitting to a node at distance D.

3.3 Inter-Node Distances

In the following, we consider a binomial point process [50] consisting of N nodes uniformly distributed on a disk of radius ρ. We derive the distributions of the distances between these nodes to a node of interest which is located outside of the disk at a distance D from the center (see Fig. 3.1). The expected values of these distances will then be used to predict the performance of multistage cooperative broadcast in Sec. 3.4.

3.3.1 Inter-Node Distance Distributions

In the sequel, we apply the approach of [50] to determine the distance distribution between the n-th closest node and an arbitrary node of interest to our specific geometry (Fig. 3.1). To this end, we first determine the probability of a node being located in a circle of radius r around the node of interest. Therewith, we conclude on the cumulative distribution function (CDF) of the n-th closest node and finally determine the corresponding PDF.

The probability that a node is located in a circle of radius r around the node of interest is given as the area of the intersection of this circle with the disk divided by the total area of the disk. The area A of the intersection can be found as the sum of the area of two segments as sketched in Fig. 3.1

$$
\begin{aligned}
A &= A_1 + A_2 \\
&= \frac{1}{2}\rho^2 \left(2\beta - \sin(2\beta)\right) + \frac{1}{2} r^2 \left(2\alpha - \sin(2\alpha)\right)
\end{aligned} \tag{3.2}
$$

with

$$
\beta = \arccos\left(\frac{\rho^2 + D^2 - r^2}{2\rho D}\right) \tag{3.3}
$$

and

$$
\alpha = \arccos\left(\frac{r^2 + D^2 - \rho^2}{2rD}\right) \tag{3.4}
$$

for $r \in [D-\rho, D+\rho]$, the interval of interest. Hence, the probability that a specific node is located in a circle of radius r around the node of interest is given as

$$
p = \frac{\rho^2 \left(2\beta - \sin(2\beta)\right) + r^2 \left(2\alpha - \sin(2\alpha)\right)}{2\pi\rho^2} \tag{3.5}
$$

for $r \in [D-\rho, D+\rho]$. For $r < D-\rho$ the probability is $p = 0$ and for $r > D+\rho$ it is $p = 1$. From the binomial distribution, the probability that less than n of the N nodes on the disk are within the circle of radius r is then given by

$$
\bar{F}_n(r) = \sum_{k=0}^{n-1} \binom{N}{k} p^k (1-p)^{N-k}. \tag{3.6}
$$

At the same time, $\bar{F}_n(r)$ corresponds to the probability that the n-th closest node is at distance r or more. Hence, the CDF of the distance of the n-th closest node is given by $F_n(r) = 1 - \bar{F}_n(r)$. The corresponding PDF can then be found by taking the derivative of $F_n(r)$ with respect to r as given in [50] as

$$
f_n(r) = \frac{-d\bar{F}_n(r)}{dr} = \frac{dp}{dr} \frac{(1-p)^{N-n} p^{n-1}}{B(N-n+1, n)}, \tag{3.7}
$$

where

$$
B(a,b) = \int_0^1 t^{(a-1)} (1-t)^{(b-1)} dt \tag{3.8}
$$

is the beta function [53]. The derivative of p can be found as

$$
\frac{dp}{dr} = \frac{2r}{\pi\rho^2} \left(\frac{\rho}{D}\sqrt{1-b^2} - \frac{r}{D}\sqrt{1-a^2} + \arccos(a)\right), \tag{3.9}
$$

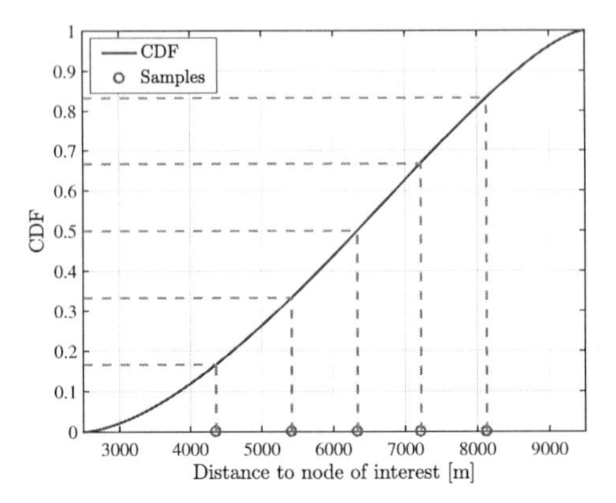

Figure 3.2.: CDF of the distance of an arbitrary node on the disk to the node of interest for $\rho = 3500$ m, $D = 6000$ m and the approximations of the expected values of the ordered distances for $N = 5$ (red circles).

with

$$a = \frac{r^2 + D^2 - \rho^2}{2rD}, \quad b = \frac{\rho^2 + D^2 - r^2}{2\rho D}. \tag{3.10}$$

3.3.2 Expected Values of Inter-Node Distances

For the performance prediction of the cooperative broadcast, which will be presented in Sec. 3.4, only the expected values of the distances $\mathsf{E}[d_n]$ are required. Unfortunately, we are not aware of an analytical solution to $\mathsf{E}[d_n] = \int f_n(r) r dr$ for the PDFs given in Eq. (3.7). Thus, the expected values would have to be evaluated numerically. As this requires high computational effort, we discuss an approximation in the following. It allows to determine the expected values of the distances much faster at high accuracy. The derived PDFs in Eq. (3.7), in the end, are only used to verify the approximations by numerical integration, as shown in Fig. 3.3.

For the approximation of the expected values $\mathsf{E}[d_n]$ we make use of the fact that the distribution of the CDF values of the ordered distances is well known. Namely, according to the probability integral transform, for a random variable X distributed according to $f_X(x)$ with CDF $F_X(x)$, the CDF value of random variable X is uniformly distributed on the interval $[0, 1]$, $Y = F_X(X) \sim \mathcal{U}(0, 1)$. Considering now N i.i.d. random variables $X_n \sim f_X(x) \; \forall \, n \in \{1, \dots, N\}$, ordered such that $x_1 \leq x_2 \leq \cdots \leq x_N$,

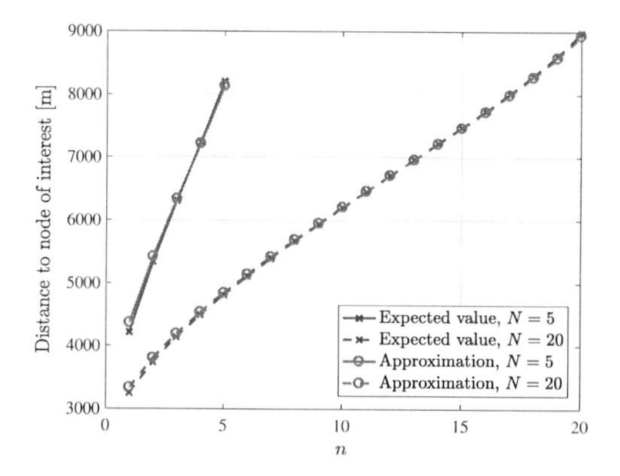

Figure 3.3.: Numerically evaluated expected values of the distance of the n-th closest node to the node of interest and their approximations for $\rho = 3500$ m, $D = 6000$ m and $N \in \{5, 20\}$.

the corresponding CDF values y_n are distributed according to the beta distribution [53], $Y_n = F_X(X_n) \sim \text{Beta}(n, N+1-n)$, with mean $\mu_n = n/(N+1)$ and variance

$$\sigma_n^2 = \frac{n^2 + n(N+1)}{(N+1)^2(N+2)}. \tag{3.11}$$

That is, the expected value of the random variable $Y_n = F_X(X_n)$ is given as $\mathsf{E}[Y_n] = \mu_n$.

In order to approximate $\mathsf{E}[d_n]$ we now determine the CDF of the distance of a single node uniformly distributed on the disk to the node of interest, denoted by $C(r)$. We then consider N nodes i.i.d. according to this distribution, order their distances such that $d_1 \leq d_2 \leq \cdots \leq d_N$ and evaluate $\mathsf{E}[d_n] = \mathsf{E}[C^{-1}(U_n)]$ with $U_n = C(d_n)$. The CDF $C(r)$ is given by the normalized covered area A in Eq. (3.2) as

$$C(r) = \frac{\rho^2 \left(2\beta - \sin(2\beta)\right) + r^2 \left(2\alpha - \sin(2\alpha)\right)}{2\pi\rho^2} \tag{3.12}$$

for $r \in [D-\rho, D+\rho]$. For $0 \leq r < D-\rho$ it is given as $C(r) = 0$ and for $r > D+\rho$ as $C(r) = 1$. It is shown in blue in Fig. 3.2 for $\rho = 3500$ m and $D = 6000$ m.

Following [53] by considering $Q = C^{-1}$, expanding $Q(U_n)$ in a Taylor series around

μ_n, and taking the expectation of it, we get

$$\begin{aligned}\mathsf{E}[d_n] =& Q(\mu_n) + \mathsf{E}[(U_n - \mu_n)]Q'(\mu_n) + \frac{1}{2}\mathsf{E}[(U_n - \mu_n)^2]Q''(\mu_n) + \ldots \\ =& Q(\mu_n) + \frac{n^2 + n(N+1)}{2(N+1)^2(N+2)}Q''(\mu_n) + \ldots .\end{aligned} \tag{3.13}$$

As $\mathsf{E}[(U_n - \mu_n)]$ evaluates to 0 ($\mathsf{E}[(U_n)] = \mu_n$), the term of $Q'(\mu_n)$ vanishes. By neglecting the higher order derivatives, we end up with

$$\mathsf{E}[d_n] \approx Q(\mu_n) = C^{-1}\left(\frac{n}{N+1}\right) \tag{3.14}$$

as a straightforward approximation of the expected values of the ordered distances. This procedure of inversely sampling the CDF is illustrated in green with the resulting samples in red for $N = 5$ in Fig. 3.2.

As the slope of the considered CDF is not changing strongly and is almost constant on a large interval, its higher order derivatives are rather small. Hence, the contribution of the higher order terms in the Taylor series in Eq. (3.13) are small. Furthermore, the term in front of $Q''(\mu_n)$ in Eq. (3.13) scales at least with $1/N$ (depending on n). The accuracy of the approximation is thus expected to increase for higher N. A numerical evaluation of $\mathsf{E}[d_n] = \int f_n(r) r dr$ and the corresponding approximations are shown in Fig. 3.3 for $\rho = 3500$ m and $D = 6000$ m. A value of $N = 5$ (very low node density) already yields high accuracy. Only the approximation of the distance to the closest node is slightly off the real value. For $N = 20$, the approximation is very accurate for all nodes.

As we are not aware of an analytical solution of the inverse of the CDF, C^{-1}, we evaluate the values of the r corresponding to $C(r) = \frac{1}{N+1}, \frac{2}{N+1}, \ldots, \frac{N}{N+1}$ numerically. This can be done efficiently by a line search on $r \in [D - \rho, D + \rho]$.

3.3.3 Approximation of Inter-Node Distances

For a large number of nodes N on the disk, their actual ordered distances $d_1 \leq d_2 \leq \cdots \leq d_N$ to the node of interest can be closely approximated by their expected values $\mathsf{E}[d_n]$. This can be seen by considering the variance of the corresponding CDF values $U_n = C(d_n)$ given in Eq. (3.11). It decreases with at least $1/N$. As the CDF $C(r)$ is locally almost linear, this directly translates to the variance of $d_n = C^{-1}(U_n)$. That is, the variance of d_n is also strongly decreasing for increasing N. Thus d_n can be accurately approximated by its expected value for large N.

3.4 Performance Prediction of Multistage Cooperative Broadcast

For the first hop of multistage cooperative broadcast it is simple to find the communication range and the PDF of the number of covered nodes analytically. However, it becomes very hard from the second hop on, as the distance which can be overcome in one hop depends on the number of nodes reached in the previous hop (Poisson distributed random variable) as well as on their spatial distribution. In this context, we present a low complexity performance prediction of the multistage cooperative broadcast in the following.

Analogously to Monte Carlo evaluations, we consider each hop separately and determine in every direction the coverage range and the nodes reached. Based on these results, the next hop is evaluated likewise and so on until the required area is covered, respectively all required nodes are reached. However, instead of averaging over many realizations as in the Monte Carlo evaluations, we employ the expected values of the coverage distance and the number of nodes reached in each hop (requiring only low computational complexity). To this end, we make the following assumptions for each hop:

- The coverage area contains the expected number of nodes.
- These nodes are distributed according to a binomial point process.
- The message spreads circularly.

Hence, the expected coverage range can be determined from the distance distributions respectively its expected values derived in Sec. 3.3. Due to the circular spreading, it is the same in all directions.

Intuitively, these steps can be repeated iteratively in each hop. Thereby, the assumptions and simplifications made are motivated as follows. If the locations of the nodes form a Poisson point process in $\mathbb{R}^2$, the nodes contained in any subset of the plane are uniformly distributed in this subset. Hence, for a sufficiently large node density the message spreads circularly as from all directions the distribution of the distances from a node of interest outside of the disk to the nodes inside look alike. Therefore, also the communication range in all directions is approximately the same. For a high node density, the circular spreading of a message in multistage cooperative broadcast was also shown in [21]. Furthermore, for a large number of nodes in the disk, the distances to these nodes are well approximated by the expected values of the distances as discussed in Sec. 3.3.3. If the coverage distance is accurately approximated also the covered number of nodes is accurately approximated by its expected value. Hence, for high node densities, each hop can be considered separately, based on the coverage area of the previous hop.

The performance prediction of the cooperative broadcast is summarized in the following and its performance is evaluated. We thereby show that the stated assumptions are also valid for low node densities and allow to accurately predict the average performance of the multistage cooperative broadcast.

3.4.1 First hop

As we consider a fixed transmission rate R_{min}, the distance which can be overcome in the first hop, $d_{\text{max},1}$, is also fixed and can be found from the achievable rate

$$R_{\text{min}} = \log_2\left(1 + \frac{P_{\text{Tx}} \cdot |h|^2 \cdot d_{\text{max},1}^{-\gamma}}{\sigma_w^2}\right) \tag{3.15}$$

as

$$d_{\text{max},1} = \left(\frac{P_{\text{Tx}}}{(2^{R_{\text{min}}} - 1) \cdot \sigma_w^2}\right)^{1/\gamma}, \tag{3.16}$$

as $|h|^2 = 1$.

3.4.2 Further hops

From the second hop on, we consider that the number of nodes reached so far (denoted by N_{k-1}) is given by its expected value. Furthermore, we assume that these nodes are uniformly distributed on a disk of radius $d_{\text{max},k-1}$ centered around the origin, where k denotes the hop index.

The expected number of covered nodes (including the source) is given by

$$\mathsf{E}[N_{k-1}] \geq \pi \cdot \mathsf{E}[d_{\text{max},k-1}]^2 \cdot \delta, \tag{3.17}$$

where the lower bound follows from Jensen's inequality (for $k = 2$ it is fulfilled with equality as $d_{\text{max},1}$ is deterministic). As an integer number of nodes is required to determine the coverage range of the next hop, we consider the ceiling of this value as the number of covered nodes (due to the $\geq$ in Eq. (3.17)), i.e.

$$N_{k-1} = \lceil \pi \cdot \mathsf{E}[d_{\text{max},k-1}]^2 \cdot \delta \rceil . \tag{3.18}$$

If all these nodes are simultaneously transmitting the same message with a different codebook, the expected achievable rate of a node of interest outside of this disk at distance D from the origin is given by Jensen's inequality as

$$\mathsf{E}[R] \leq \log_2 \left(1 + \frac{P_{\text{Tx}} \sum_{n=1}^{N_{k-1}} \mathsf{E}[d_n^{-\gamma}]}{\sigma_w^2} \right) , \tag{3.19}$$

where d_n are the distances to the N_{k-1} nodes inside the disk. The minimum required received signal power to achieve R_{min} can then be approximated by

$$P_{\text{min}} = P_{\text{Tx}} \cdot \sum_{n=1}^{N_{k-1}} \mathsf{E}[d_n]^{-\gamma} \approx (2^{R_{\text{min}}} - 1) \cdot \sigma_w^2 . \tag{3.20}$$

The approximation $\approx$ comes from the fact that from Jensen's inequality $\mathsf{E}[d_n^{-\gamma}] \geq \mathsf{E}[d_n]^{-\gamma}$. As this inequality is in the opposite direction of the inequality in Eq. (3.19), we can't conclude on a bound based on the expected values of d_n.

Eq. (3.20) is key to estimate the distance $d_{\text{max},k}$ which can be overcome in this hop. In Sec. 3.3 we have determined the expected values of the sorted distances $\mathsf{E}[d_n]$ of all N_{k-1} nodes within a disk to a node of interest outside this disk. These expectations are a function of the disk radius ρ (Fig. 3.1) and the distance D between the node of interest and the center of the disk. Thus we can estimate the maximum range achieved in this hop as the value of D for which Eq. (3.20) is just fulfilled. Due to the difficulty to find the distance D analytically (would require the inverse of $C(r)$), it is determined by a simple line search.

3.4.3 Performance Evaluation

The performance of the multistage cooperative broadcast can now be predicted for large number of nodes and hops by iteratively applying the steps described above until the required average number of nodes is reached or the required average distance is

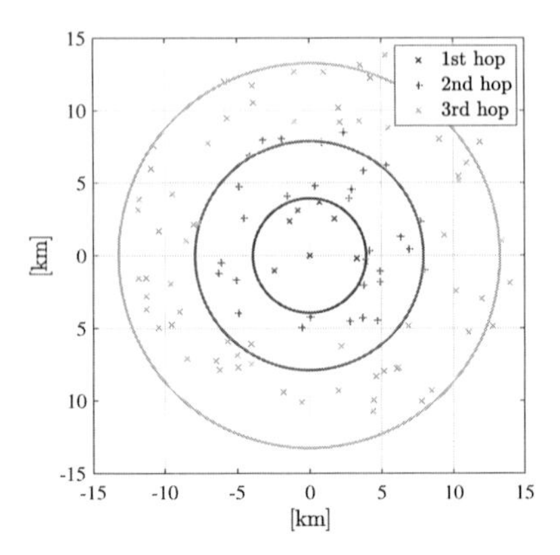

Figure 3.4.: Example of covered nodes per hop for a network with $\delta = 0.2$ nodes/km^2 (crosses) and the prediction of the coverage area (solid circles).

overcome. The following results show the predicted performance compared to Monte Carlo simulations over 4 hops. The parameters have been set to $R_{\text{min}} = 1$ bps/Hz, $\sigma_w^2 = 2.62 \cdot 10^{-15}$ W, $P_{\text{Tx}} = 0.01$ W and $\gamma = 3.5$ with node densities of $\delta \in \{0.1, 0.2, \ldots, 1\}$ nodes/km^2. These numbers are motivated by a military MANET with typically large dimensions and low data rates (see Chap. 7). The resulting coverage range of the first hop is thus given as $d_{\text{max},1} = 3933$ m. Note that a node density of $\delta = 0.1$ nodes/km^2 is very low and corresponds to an expected value of only 5 covered nodes in the first hop.
The figures Fig. 3.4 and Fig. 3.5 show two examples of the prediction of the coverage per hop (solid circles) and the actual nodes reached in each hop for one initialization of the Monte Carlo simulations (crosses). We show the first 3 hops and consider $\delta = 0.2$ nodes/km^2 as well as $\delta = 1$ node/km^2. It can be observed that although the prediction of the coverage area is already quite accurate for $\delta = 0.2$ nodes/km^2 (Fig. 3.4), there are still nodes significantly outside of the predicted area. The coverage area is only roughly approximated by a disk due to the heavy influence of the actual node topology. For $\delta = 1$ node/km^2 (Fig. 3.5), this effect is already strongly reduced and the coverage area is well approximated by a disk with radius given by the predicted coverage distance.

The strong influence of the random node topology for low node densities can also be seen in Fig. 3.6, which exemplarily shows the empirical CDFs of the coverage distance in the second hop of the Monte Carlo simulations for a selection of node densities. We consider the coverage distance for each random node topology of the Monte Carlo

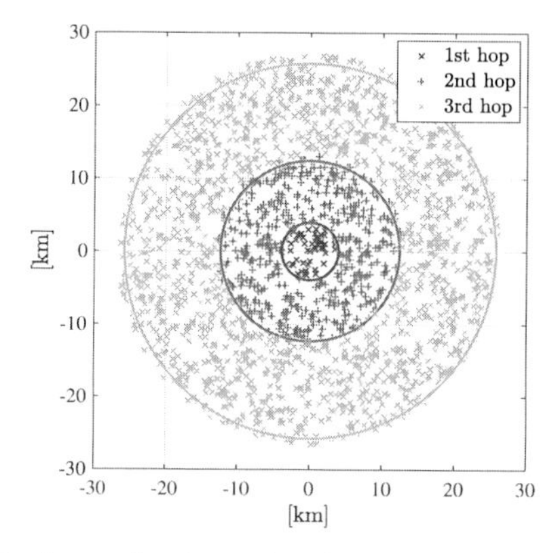

Figure 3.5.: Example of covered nodes per hop for a network with $\delta = 1$ nodes/km^2 (crosses) and the prediction of the coverage area (solid circles).

simulations as the maximal range achieved in the x-direction[1]. While the variance for low node densities is rather high, it is reduced for increased node density. The higher δ, the more similar the d_n look like for nodes outside of the coverage area as the variance in the d_n decreases (see Sec. 3.3.3), and thus to lower the variance in the coverage distance. Note that for $\delta = 0.1$ nodes/km^2, 6 out of 1000 Monte Carlo simulations resulted in an outage in the second hop. For higher node densities, no outages could be observed.

Despite the strong influence of the random node topology, the average distance reached in each hop is very well predicted by our approach. This is evidenced by Fig. 3.7 which compares the results for two, three and four hops as a function of the node density δ. Only for very low node densities a slight offset of the prediction compared to the average values of the Monte Carlo simulations can be observed. For increasing node densities, the prediction matches the average values very accurately. Even over 4 hops, the accuracy is still very high. That is, the randomness of the first 3 hops does not affect the prediction accuracy of the fourth hop.

As is to be expected by the accurate results on the average coverage distance, we also get very precise results for the number of nodes reached per hop. The corresponding predictions and averaged simulation values are shown in Fig. 3.8. Nevertheless,

[1] For a given node topology the coverage distance depends on the direction considered. However, for symmetry reasons the average coverage distance is independent of the direction.

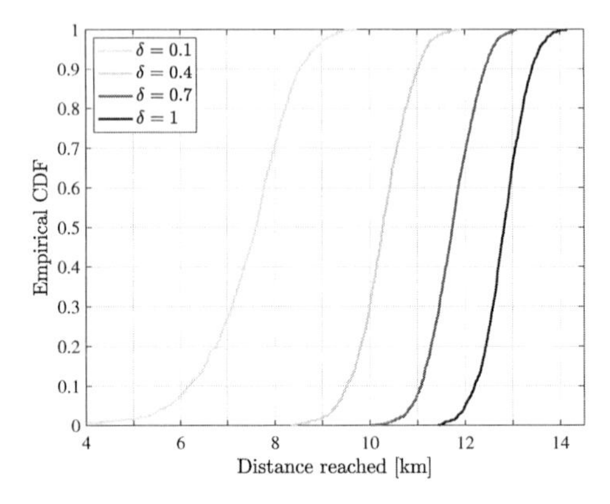

Figure 3.6.: Empirical CDF of the distance reached in the second hop of the Monte Carlo simulations for multistage cooperative broadcast. The resulting variances in the distances reached are 0.89 km^2 for $\delta = 0.1$ nodes/km^2, 0.33 km^2 for $\delta = 0.4$ nodes/km^2, 0.27 km^2 for $\delta = 0.7$ nodes/km^2 and 0.21 km^2 for $\delta = 1$ node/km^2

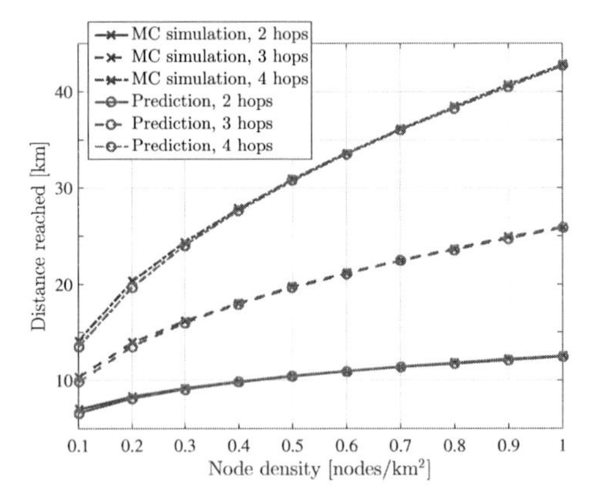

Figure 3.7.: Average distance reached per hop in multistage cooperative broadcast: prediction compared to Monte Carlo simulations.

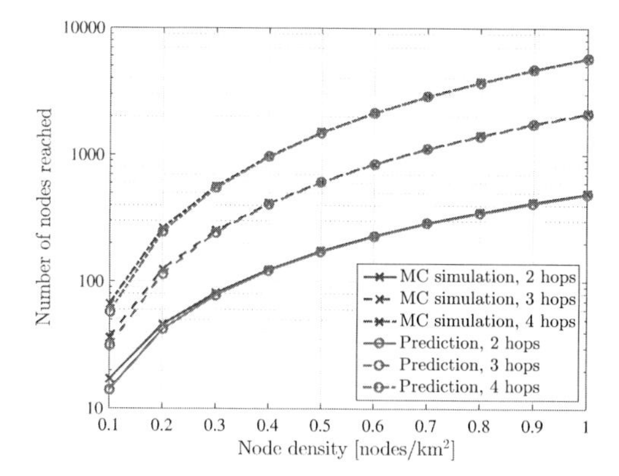

Figure 3.8.: Average number of nodes reached per hop in the multistage cooperative broadcast: prediction compared to Monte Carlo simulations.

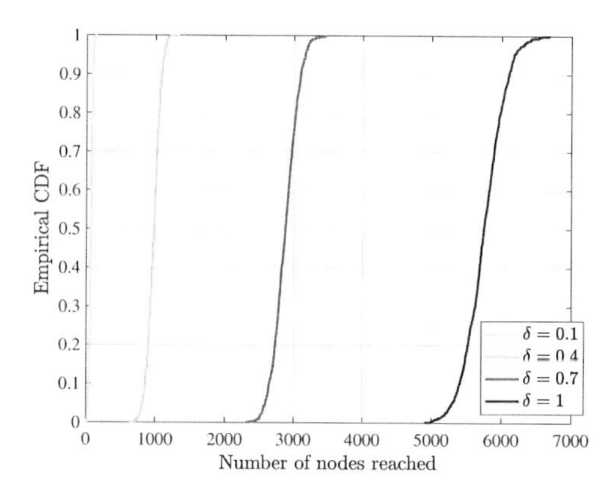

Figure 3.9.: Empirical CDF of the number of nodes reached in the fourth hop of the Monte Carlo simulations for multistage cooperative broadcast.

although the average values are very accurate, the variance is quite high. This can be seen in Fig. 3.9 which shows the empirical CDF of the number of nodes reached in the fourth hop for a selection of node densities. Especially for high node densities (achieving large coverage ranges), the variance is rather big. This comes from the fact that the number of covered nodes changes quadratically with the coverage distance. Hence, already small variations in the coverage range lead to large variations in the number of nodes.

3.4.4 Validity of Results

While the presented results are very promising, a reasonable question is how generally valid these results are with changing parameters and increasing number of hops. Furthermore, the validity of the performance prediction for a setup with small scale fading is of great interest.

As was shown for the setup above, the accuracy with increasing number of hops is not decreasing until 4 hops. For higher number of hops no Monte Carlo simulations are available for comparison due to the computational complexity for the huge number of nodes at high node densities (see Fig. 3.8). However, the accuracy is not expected to decrease for higher number of hops, as the number of nodes grows faster than the coverage radius (see Fig. 3.7 and Fig. 3.8). Hence, the approximation of the expected values is more precise at higher hop counts (see Sec. 3.3.2), and the expected values approximate the real distance more accurately (see Sec. 3.3.3).

Considering the network parameters, the prediction accuracy is expected to decrease when only a very low expected number of new nodes is covered in each hop. This occurs for very low node densities, or also for high R_{min}, a high path loss or low P_{Tx} at low node densities. Nevertheless, in these operating regimes the performance of the cooperative broadcast is anyways strongly affected by outages (i.e. in a certain hop no new nodes can be reached and thus the message can not be delivered to all intended nodes), and is therefore not in the focus of our performance prediction.

In a setup with small scale fading (e.g., Rayleigh fading with $h_i \sim \mathcal{CN}(0,1)$ and thus $\mathsf{E}[|h_i|^2] = 1$, as we will consider it in the following) more randomness is added to the system. While a node at distance D_1 might be unable to decode the message due to its poor channel to the transmitters, a node at distance $D_2 > D_1$ might be able to decode at the same time. Hence, no fixed coverage range can be assigned for a given node topology. To still get an impression on the validity of the performance

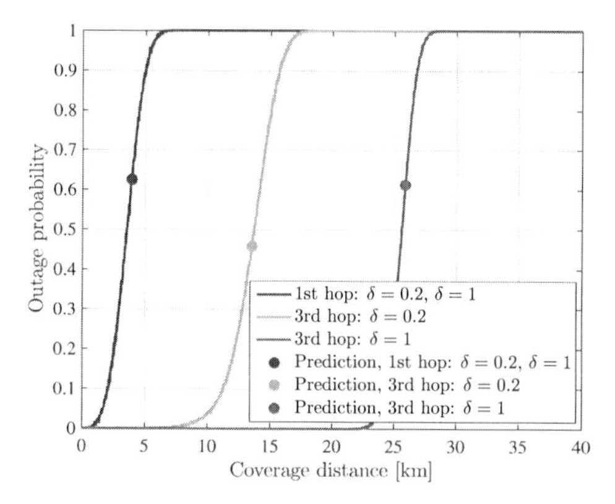

Figure 3.10.: Outage probability in the multistage cooperative broadcast in dependency of the coverage distance for $\delta = 0.2$ nodes/km^2 and $\delta = 1$ node/km^2 for 1 and 3 hops, and the corresponding prediction of the coverage distance.

prediction for such setups, we evaluate the outage probability for virtual nodes on the x-axis with Monte Carlo simulations. That is, for each node topology, we place virtual nodes on the x-axis and draw fading coefficients for them to all real nodes. We then evaluate the received signal power in each hop and check whether R_{min} can be achieved or not. By evaluating a large number of network realizations the outage probability can be determined for a given distance in a certain number of hops. These outage probabilities are shown in Fig. 3.10 for $\delta = 0.2$ nodes/km^2 and $\delta = 1$ node/km^2 for the first hop and the third hop.

As can be seen, the performance predictions (indicated by the circles in Fig. 3.10) lead to roughly 62 percent outage probability in the first hop and 46 percent for $\delta = 0.2$ nodes/km^2 respectively 61 percent for $\delta = 1$ node/km^2 in the third hop. Such values are reasonable as we consider the average range in the performance prediction. Even without small scale fading similar values would occur (see Fig. 3.6). For increasing number of hops the outage probability for the predicted coverage distance is decreasing. For the high node density ($\delta = 1$ node/km^2), this comes from the large number of transmitting nodes and thus the strongly increased diversity in the received signal. That is, while the accuracy of the approximation of the actual distances by their expected values is increasing (see Sec. 3.3.3), the impact of a single fading realization decreases due to the large number of transmitters. Hence, the achievable rate is well approximated by Eq. (3.19). The good approximation can also be seen in the decreased variance of the outage probability curve (compared to the first hop).

In contrast to that, the variance of the outage probability curve in the third hop for the low node density ($\delta = 0.2$ nodes/km^2) is bigger and the outage probability of the predicted coverage distance is decreased. This is due to the lower diversity in the received signal. Extreme cases in the fading realizations have a bigger impact in the received signal power leading to the larger variance. Furthermore, a very strong channel of one of the close nodes to the destination can have a high impact on the received signal power and can thus strongly increase the transmission range. At the same time the impact of very weak channel is smaller. Hence, the outage probability of the predicted coverage distance is decreased.

In summary, the presented performance prediction leads to very reasonable results even for a system with Rayleigh fading, which is a very pessimistic assumption as it considers no line-of-sight component at all. However, as the performance prediction only provides an average coverage range, the outage probability has to be kept in mind when results are interpreted.

3.5 Multi Hop Transmission Performance Prediction

Similar to multistage cooperative broadcast, the derived distance distributions respectively their expected values can also be used to predict the performance of the multi hop transmission, where only one node at the time forwards a message from the source to the destination. Again, we consider time slots of fixed length and a transmission rate per hop of $R_{\min}$ bps/Hz.

Analogously to the multistage cooperative broadcast, we consider each hop separately and determine the expected coverage range and the expected number of nodes which could be reached in the corresponding number of hops. The steps are very similar, with the difference that we consider only one node transmitting at a time for all hops.

3.5.1 First Hop

The first hop is equivalent to multistage cooperative broadcast. The coverage range of the source is given as

$$d_{\max,1} = \left(\frac{P_{\mathrm{Tx}}}{(2^{R_{\min}} - 1) \cdot \sigma_w^2}\right)^{1/\gamma}. \tag{3.21}$$

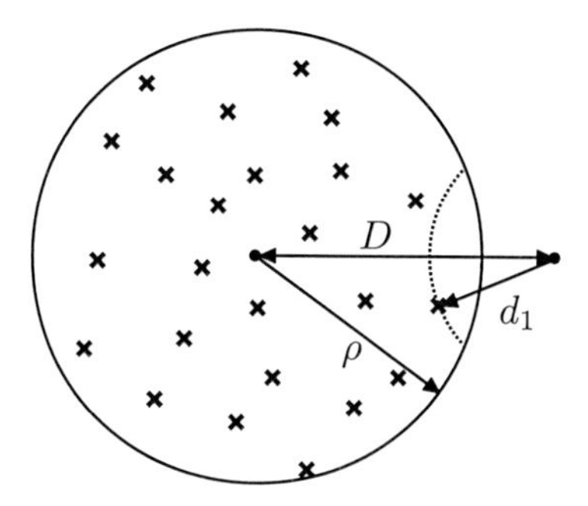

Figure 3.11.: Considered setup for the multi hop transmission: N_{k-1} nodes distributed on a disk of radius $\rho = d_{\max,k-1}$ and the node of interest at distance D.

That is, all nodes within this range can be reached in one hop.

3.5.2 Further Hops

From the second hop on, we make the same assumptions as for the performance prediction of multistage cooperative broadcast.

- The coverage area containing the nodes which could be reached in a certain number of hops is given by a circle with the coverage range as radius.
- The coverage area contains the expected number of nodes.
- These nodes are distributed according to a binomial point process.

To determine the coverage range in hop k, we consider a setup as shown in Fig. 3.11, where the N_{k-1} nodes which can be reached within $k-1$ hops are uniformly distributed on a disk of radius $\rho = d_{\max,k-1}$. From the perspective of the node at distance D, the closest node inside the disk would lead to the maximal coverage range in its direction. That is, the node at distance D can be reached within k hops if the distance to the closest node inside the disk is smaller than the coverage radius of a single node, i.e. $d_1 \leq d_{\max,1}$. For a sufficiently large number of nodes inside the disk, this distance is approximately given by $\mathsf{E}[d_1]$ and can be determined from the distance distributions as derived in Sec. 3.3. Hence, to determine the new coverage radius, we look for D such that $\mathsf{E}[d_1] = d_{\max,1}$. As the problem is symmetric in all directions, any node inside the coverage range $d_{\max,k} = D$ is considered to be reachable within k hops.

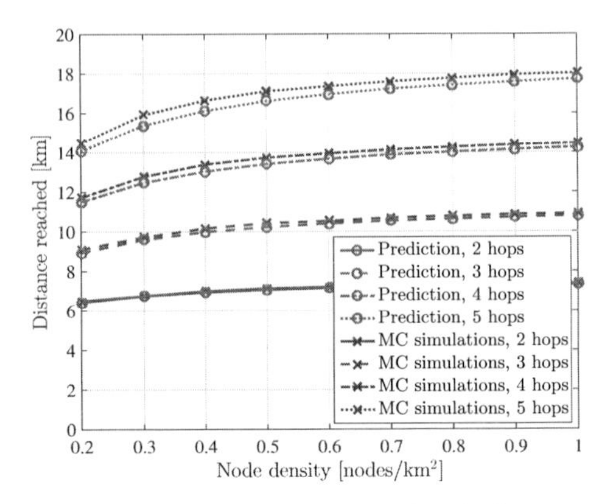

Figure 3.12.: Average distance reached with classical multi hop transmission within indicated number of hops: prediction compared to Monte Carlo simulations.

3.5.3 Performance Evaluation

Analogously to the multistage cooperative broadcast, the performance of multi hop transmission can be predicted by iteratively applying the steps described above. For the performance evaluation we compare the predicted performance to the average value of Monte Carlo simulations. Again, we consider the coverage distance for each random node topology of the Monte Carlo simulations as the maximum range achieved in x-direction. The system parameters are set equally as for the evaluation of the multistage cooperative broadcast in Sec. 3.4.3, except that we do not evaluate the lowest node density due to the large number of outages in the Monte Carlo simulations: $R_{\min} =$ 1 bps/Hz, $\sigma_w^2 = 2.62 \cdot 10^{-15}$ W, $P_{\text{Tx}} = 0.01$ W, $\gamma = 3.5$ and $\delta \in \{0.2, 0.3, \ldots, 1\}$ nodes/km^2.

Fig. 3.12 shows the resulting average distances which can be overcome in up to 5 hops in dependence of the node density. It can be seen that for 2 hops, the average distance reached can be predicted very accurately, even for low node densities. For an increasing hop count the prediction gets less accurate. The reduced accuracy comes from the strong influence of the random node topology. As only one node is transmitting, the derivations of the real distance from the expected value of the distance impact the results much stronger than in the cooperative broadcast, where the individual derivations of all transmitting nodes partially compensate each other. For increasing hop counts, the impact of this randomness is increasing. Still, even for 5 hops, the

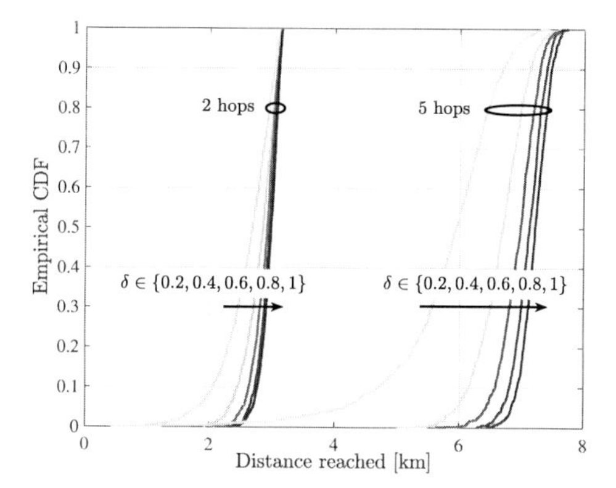

Figure 3.13.: Empirical CDF of the distance reached in the second and fifth hop of the Monte Carlo simulations for multi hop transmission.

approximation of the distance which can be overcome is quite accurate.

The increasing impact of the randomness can also be observed in Fig. 3.13. It shows the empirical CDF of the distance which can be reached in the Monte Carlo simulations for 2 and 5 hops and selected node densities. While for two hops the variance in the results is rather small, it strongly increases for 5 hops, especially for low node densities. Obviously for higher node densities, the variance in the results is decreasing, as it is more likely to have the closest node at the distance as predicted. This also increases the accuracy in the prediction (slightly visible in Fig. 3.12).

3.5.4 Validity of Results

Analogously to the performance prediction of multistage cooperative broadcast, we shortly discuss the validity of the results of multi hop transmission for varying number of hops and node densities, as well as for a setup with small scale fading.

As indicated by the results in Fig. 3.12, the accuracy of the prediction is expected to further decrease for higher hop counts due to the additional randomness with each hop. For increasing node density the accuracy is generally increasing, as the distance to the closest node can be approximated more accurately. For lower node densities than the ones shown, the accuracy is decreasing. However, networks with lower node density are anyways dominated by outages (i.e. the multi hop transmission fails due

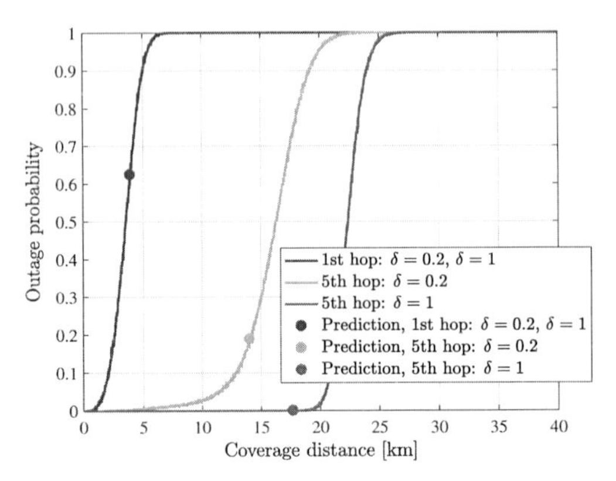

Figure 3.14.: Outage probability in the multi hop transmission in dependency of the coverage distance for $\delta = 0.2$ nodes/km^2 and $\delta = 1$ node/km^2 for 1 and 5 hops, and the corresponding prediction of the coverage distance.

to missing forwarding nodes).

Considering a setup with small scale fading (again Rayleigh fading is considered) the same evaluations are done as for the multistage cooperative broadcast. The corresponding outage probabilities are shown in Fig. 3.14 for $\delta = 0.2$ nodes/km^2 and $\delta = 1$ node/km^2 for the first hop and the fifth hop. Considering the variance of the outage probability curves, similar observations can be made as for the multistage cooperative broadcast. Although only one node is transmitting at a time in multi hop transmission, there is still a large diversity in the network, as a message can take various paths to the destination. The higher the node density, the higher the diversity due to the larger number of nodes. Hence, the variance in the outage probability curve decreases for increasing density. Different to the performance prediction of the multistage cooperative broadcast, the outage probability of the predicted coverage distances are strongly decreasing for increasing number of hops as well as for increasing node density. This is also due to the increased diversity in the system. As only one node is transmitting at a time, the impact of the channel realizations is very big. For a larger number of nodes, the probability that a path with lots of strong channels can be found is increasing. Hence, the transmission ranges are increasing and the performance prediction is underestimating the coverage range.

Hence, for multi hop transmission with Rayleigh fading channels, the performance prediction rather underestimates the real achievable distance. This underestimation is increasing for increasing node density and increasing number of hops. However, for setups with possible line-of-sight connections to close nodes, the accuracy of the performance prediction is expected to be higher.

3.6 Conclusion

In this chapter, we presented a novel performance prediction of the multistage cooperative broadcast and the multi hop transmission based on the expected values of the inter-node distances. We derived the required distance distributions, discussed an accurate and computationally simple approximation of its expected values and evaluated its performance in comparison to extensive Monte Carlo simulations. It was shown that the proposed approach leads to very accurate average results at low computational complexity. Hence, it can be applied to predict the performance for large networks in reasonable time.

4 Leakage Based Multi-User MIMO Precoding

In this chapter, we investigate the relation between transmit power and leakage power in leakage based precoding and derive low complexity solutions for rate optimal precoding under leakage and transmit power constraints, as well as for leakage based target rate precoding.

In order to exploit the spatial multiplexing gain and serve multiple nodes simultaneously by a multi-antenna transmitter, a sophisticated multi-user multiple-input multiple-output (MIMO) precoding scheme is necessary to reduce the co-channel interference among the users. Such a multi-antenna transmitter could either be a node with multiple antennas, such as a base station in a cellular network or an access point in a wireless local area network (WLAN), or equivalently a virtual antenna array formed by a cluster of cooperating nodes in a mobile ad hoc network (MANET). Multi-user MIMO precoding schemes have been studied extensively in the context of coordinated multipoint transmission and various proposals have been made [16]. A very promising approach is leakage based precoding, where for a desired signal to an intended user the co-channel interference is reduced by minimizing the transmitted interference to all unintended users (further called leakage) [18, 54–57]. This leads to a decoupled optimization problem for the precoding design of each user, requires only limited channel state information, and a closed-form solution can be found (in contrast to, e.g., maximizing the signal to interference and noise ratio for each user [58]). Furthermore, there is no constraint on the number of transmit or receive antennas in the system (as, e.g., for zero-forcing [17]), and it is well suited for the application in cellular networks combined with coordinated multipoint transmission [59].

The aforementioned leakage based precoding schemes minimize the leakage power produced by the desired signal and transmit at full power. However, this can not guarantee that a maximal leakage level is not exceeded. Hence, although the total leakage is minimized, the interference at some nodes in the network can be very high, severely decreasing their performance. Furthermore, when the generated leakage for a certain link is very small, there is no point in further reducing it, as the benefit for the other nodes is fairly small. In such a case, the focus is rather put on the desired signal instead.

In this context, we study the interdependency between leakage reduction and transmit power control for rate maximization in leakage based precoding for a multi-user MIMO scenario. We propose a leakage based precoding scheme which provides the rate optimal precoding under a joint leakage and transmit power constraint and derive an iterative closed form solution for it. Any leakage level can be achieved between the egoistic case, where the leakage is not reduced at all, and the altruistic case, where the leakage is minimized. This allows to optimize the throughput of a multi-user MIMO setup by carefully choosing the leakage power and transmit power constraints.

As multi-user MIMO precoding often leads to highly unequally distributed achievable rates among the users, we furthermore present a *target rate precoding* (i.e. precoding with a quality of service constraint in terms of a target rate) based on the transmit power - leakage power trade-off of leakage based precoding. It allows to optimize the precoding such that a large share of the users achieve a predefined target rate in a decentralized fashion at low complexity.

For the evaluation of the proposed schemes, we focus on the application in cellular networks with multi-antenna base stations and coordinated multipoint transmission. However, all schemes can also be directly applied for transmit cooperation with a virtual antenna array in a MANET.

This chapter is mainly based on our work in [39] and [40] and is structured as follows. In Sec. 4.1 related work in leakage based precoding and target rate precoding is discussed and in Sec. 4.2 the system model is presented. In Sec. 4.3 rate optimal precoding under joint transmit and leakage power constraints is introduced and the relation between transmit power and leakage power in leakage based precoding is discussed. Based on this relation leakage based target rate precoding is introduced in Sec. 4.4. Sec. 4.5 finally concludes the chapter.

4.1 Related Work

Leakage based precoding was originally introduced in [18] by maximizing the signal-to-leakage-plus-noise ratio (SLNR). Other works independently proposed very similar or even identical approaches so SLNR precoding [60, 61]. Due to its low computational complexity and its efficiency in reducing the co-channel interference, it is a very practical method to improve the spectral efficiency. However, [18] applies equal power allocation among the sub-streams. Hence, it is not rate optimal. To this end, [54] introduces a power allocation over the SLNR values of the sub-streams, which allows to optimize the achievable rate by adaptively selecting the number of sub-streams for each link and weighting them. In [55], an iterative transmit filter design is proposed to maximize the weighted sum capacity in multi-user MIMO systems, based on a modified definition of SLNR which also includes the receiver structure, and power allocation over the sub-streams. However, in all approaches the actually generated leakage level can't be controlled. This degrades the performance for setups with very low or very high leakage power.

An approach similar to the presented rate optimal precoding under joint leakage and transmit power constraint has independently been followed in [62] as relaxed zero forcing, and solved using standard optimization tools. Different to [62], we derive an iterative closed form solution for such a problem. Furthermore, we provide a thorough analysis on the interdependency between the leakage power and the transmit power.

Considering *target rate precoding*, various approaches have been proposed in literature such as [63–65]. In [63], a transmitter design for single antenna receivers is suggested, based on Tomlinson-Harashima precoding (a non-linear precoding). It minimizes the transmit power under target rate and bit error probability constraints. In [64], the weighted sum-power minimization problem under rate constraints and the admission problem are studied. In [65], a novel gradient-based scheme is presented which minimizes the sum transmit power under per-user rate constraints. All these approaches state a global optimization problem over all users and iteratively solve it. Furthermore, all of them consider transmit power minimization under the rate constraints without constraints on the transmit power (except [64] with the admission problem). However, in real networks, global optimization is unpractical and the transmit power is generally limited. Hence, users could be in an outage, not achieving their target rate.

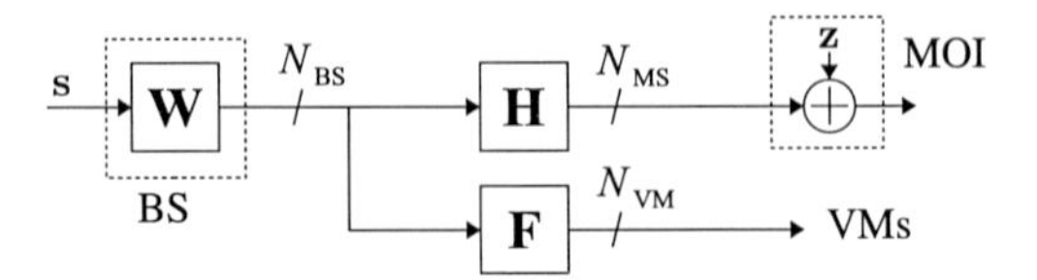

Figure 4.1.: System model.

4.2 System Model

In the following we consider a cellular setup with a base station (BS) and multiple mobile stations (MSs) which are served by the base station. The precoding is obtained for each link separately. That is, for every mobile station we consider a setup as shown in Fig. 4.1. The base station serves the mobile of interest (MOI) with the desired signal, while the other mobile stations, called victim mobiles (VMs), suffer from the generated interference. The number of antennas at the base station is N_{BS}, the number of antennas at the mobile of interest N_{MS}, and the victim mobiles are assumed to be equipped with in total N_{VM} antennas. That is, N_{MS} streams can be transmitted to the mobile of interest simultaneously (assuming $N_{\mathrm{BS}} \geq N_{\mathrm{MS}}$). Instead of a single base station, also a cluster of base stations can be considered serving the mobile of interest jointly (as in coordinated multipoint transmission). The channel from the base station to the mobile of interest is expressed by $\mathbf{H} \in \mathbb{C}^{N_{\mathrm{MS}} \times N_{\mathrm{BS}}}$ and the channel from the base station to the victim mobiles by $\mathbf{F} \in \mathbb{C}^{N_{\mathrm{VM}} \times N_{\mathrm{BS}}}$. The transmit signal vector for the mobile of interest is denoted by $\mathbf{s} \in \mathbb{C}^{N_{\mathrm{MS}}}$ with its elements $s_i \sim \mathcal{CN}(0, \sigma_{\mathrm{s},i}^2)$, $i \in \{1, \ldots, N_{\mathrm{MS}}\}$. This results in the signal covariance matrix $\mathbf{\Lambda}_{\mathrm{s}} = \mathsf{E}\left[\mathbf{s}\mathbf{s}^{\mathsf{H}}\right]$. Considering the precoding matrix $\mathbf{W} \in \mathbb{C}^{N_{\mathrm{BS}} \times N_{\mathrm{MS}}}$, the transmit covariance matrix is given as $\mathbf{Q} = \mathbf{W}\mathbf{\Lambda}_{\mathrm{s}}\mathbf{W}^{\mathsf{H}}$ and the transmit power can be found as $P_{\mathrm{Tx}} = \mathrm{tr}\left(\mathbf{Q}\right)$. At the mobile of interest we assume zero mean additive white Gaussian noise with variance σ_w^2 plus potentially interference, summarized in the term $\mathbf{z} \in \mathbb{C}^{N_{\mathrm{MS}}}$ with covariance matrix $\mathbf{K}_{\mathrm{z}}$. Hence, the received signal can be written as

$$\mathbf{r}_{\mathrm{MOI}} = \mathbf{H}\mathbf{W}\mathbf{s} + \mathbf{z}. \tag{4.1}$$

The leaked signal received at the victim mobiles is given by

$$\mathbf{r}_{\mathrm{VM}} = \mathbf{F}\mathbf{W}\mathbf{s}. \tag{4.2}$$

The leakage power can thus be written as $P_{\mathrm{L}} = \mathrm{tr}\left(\mathbf{F}\mathbf{Q}\mathbf{F}^{\mathsf{H}}\right)$.

4.3 Rate Optimal Leakage Based Precoding

In multi-user MIMO precoding all users are inherently coupled via the co-channel interference. That is, if the signal of a particular user generates a large amount of interference, the performance of the other users strongly suffers. To this end, leakage based precoding mitigates the co-channel interference by reducing the leakage power for each user to be served. Intuitively, this can be interpreted as sending a signal term with the desired message to the mobile of interest and simultaneously transmitting a compensation term. The compensation term is designed such that it does not interfere with the signal term at the mobile of interest, but minimizes the leakage to the victim mobiles. Hence, for a desired signal at the mobile of interest, the leakage power to the victim mobiles can be minimized. However, as we will show in the following, reducing the leakage power comes at the price of either increased transmit power or reduced signal power at the mobile of interest.

In fact, any leakage based precoding can be separated into the mentioned signal term $\mathbf{W}_\mathrm{s}$, serving the mobile of interest, and the compensation term $\mathbf{W}_\mathrm{c}$ to reduce the leakage power to all victim mobiles in an orthogonal subspace to $\mathbf{H}$ without disturbing the desired signal at the mobile of interest,

$$\mathbf{W} = \mathbf{W}_\mathrm{s} + \mathbf{W}_\mathrm{c}. \tag{4.3}$$

The signal term $\mathbf{W}_\mathrm{s}$ can be found as the projection of $\mathbf{W}$ onto the subspace of $\mathbf{H}$. Considering the columns of the matrix $\mathbf{U} \in \mathbb{C}^{N_\mathrm{BS} \times N_\mathrm{MS}}$ to represent an orthonormal basis of $\mathbf{H}$, this projection is given as

$$\mathbf{W}_\mathrm{s} = \mathbf{U}\mathbf{U}^\mathsf{H}\mathbf{W}. \tag{4.4}$$

$\mathbf{W}_\mathrm{c}$ is then the component of $\mathbf{W}$ orthogonal to the subspace of $\mathbf{H}$,

$$\mathbf{W}_\mathrm{c} = \mathbf{W} - \mathbf{U}\mathbf{U}^\mathsf{H}\mathbf{W}. \tag{4.5}$$

As these two terms are orthogonal to each other, we can also separate the transmit power into the power allocated to the desired signal, $P_\mathrm{s} = \mathrm{tr}\left(\mathbf{W}_\mathrm{s}\mathbf{\Lambda}_\mathrm{s}\mathbf{W}_\mathrm{s}{}^\mathsf{H}\right)$, and the power allocated to the leakage reduction, $P_\mathrm{c} = \mathrm{tr}\left(\mathbf{W}_\mathrm{c}\mathbf{\Lambda}_\mathrm{s}\mathbf{W}_\mathrm{c}{}^\mathsf{H}\right)$,

$$P_\mathrm{s} + P_\mathrm{c} = P_\mathrm{Tx}. \tag{4.6}$$

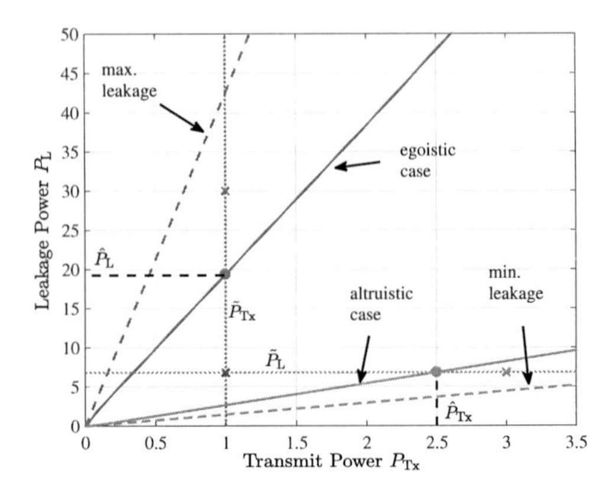

Figure 4.2.: The transmit power - leakage power plane.

That is, to reduce the leakage power, the available transmit power has to be invested partially into the compensation term, leading to a lower signal power and thus a lower rate at the mobile of interest. Hence, there is a trade-off between the achievable signal power and the resulting leakage power level.

4.3.1 The Relation of Transmit Power and Leakage Power

As we will show in the following, the trade-off between generated leakage power and the achievable rate at the mobile of interest can be steered. Depending on the amount of transmit power allocated to the compensation term, any leakage power level can be achieved between the egoistic approach where no energy at all is invested into the leakage reduction, and the altruistic approach, where the leakage power is minimized. This is illustrated in Fig. 4.2 which shows the relation of the transmit power and the generated leakage power for one specific channel realization in a setup as described in Section 4.2. The dashed red line shows the maximal achievable leakage power $P_L = \mathrm{tr}\left(\mathbf{F}\mathbf{W}\mathbf{\Lambda}_s\mathbf{W}^H\mathbf{F}^H\right)$ if transmitting with $P_{Tx} = \mathrm{tr}\left(\mathbf{W}\mathbf{\Lambda}_s\mathbf{W}^H\right)$. Points on this line are reached by precoding with the eigenvector corresponding to the maximal eigenvalue of $\mathbf{F}^H\mathbf{F}$ (i.e. only one stream is considered and $\mathbf{W}$ becomes a vector). Analogously, the dashed green line shows the minimal achievable leakage power for a given transmit power P_{Tx}. Points on this line are achieved by precoding with the eigenvector corresponding to the minimal eigenvalue of $\mathbf{F}^H\mathbf{F}$. Hence, any achievable point (P_{Tx}, P_L) lies on or in between these two lines. However, not all points in between are reasonable.

The solid red curve shows the egoistic case. That is, it shows the generated leakage power if the rate is maximized under a transmit power constraint without any measure to reduce the leakage power (no compensation term). Thus, a point on this curve corresponds to the maximum achievable rate for the respective transmit power. For any point above this curve (take the red cross as an example) there is a corresponding point (the red dot) on the curve with the same transmit power, a lower leakage power and equal or better rate. Thus, the region between the solid and dashed red curve - even though achievable - is of no practical interest. In contrast to this, the solid green curve shows the altruistic case. That is, it shows the resulting transmit power - leakage power pair if the leakage is minimized for a given rate optimal signal term. This corresponds to maximizing the rate under a leakage power constraint. That is, a point on this curve corresponds to the maximum rate achievable for the respective leakage power. For any point to the right of this curve (take the green cross as an example), there is a corresponding point (the green dot) on the green curve with the same leakage power but a lower transmit power and equal or better rate. Thus, the region between the solid and dashed green curve is as well of no practical interest.

Within the region of practical interest, the achievable rate is monotonously increasing with increasing P_{L} on a vertical line $P_{\mathrm{Tx}} = \tilde{P}_{\mathrm{Tx}}$. Intuitively, the achievable rate at the mobile of interest increases if more of the available transmit power is invested into the signal term $\mathbf{W}_{\mathrm{s}}$. At the same time, according to Eq. (4.6), less of the transmit power is available for the compensation term $\mathbf{W}_{\mathrm{c}}$, leading to a higher leakage power P_{L}. Analogously, within the region of practical interest, the achievable rate is monotonously increasing with increasing P_{Tx} on a horizontal line $P_{\mathrm{L}} = \tilde{P}_{\mathrm{L}}$. As a higher P_{Tx} is available, the power for signal as well as compensation term can be increased, leading to higher achievable rates at the mobile of interest while keeping the leakage power constant. A proof of these two statements is given in App. A.4.

4.3.2 Leakage Based Precoding Under Joint Transmit and Leakage Power Constraint

In most leakage based precoding schemes such as [18,54,60,61], there is no possibility to effectively steer the leakage power level. In [18], e.g., simply the signal-to-leakage-plus-noise ratio is maximized. While this leads to promising average results for most network setups, it is rather inefficient if either very low or very high leakage power is generated for a certain link. If the leakage level is very low, it is not worth to further decrease it. The available transmit power is rather invested into the desired signal as the other users are not really affected. If, in contrast, the generated leakage is very high, such that it even strongly hurts other users in the network after being minimized, it is better not to transmit at full power in terms of network performance.

In order to incorporate these two extremes, the precoding can be designed such that the achievable rate is maximized under a joint transmit power and leakage power constraint. That is, depending on the leakage power constraint, the transmit power is either mostly invested into the signal term, or into the compensation term. Considering only i.i.d. additive white Gaussian noise at the mobile of interest, i.e. the vector $\mathbf{z}$ is distributed according to $\mathcal{CN}(0, \mathbf{I} \cdot \sigma_w^2)$, the optimization problem can be state as

$$\begin{aligned} \max_{\mathbf{Q}} \quad & \log_2 \det \left(\mathbf{I} + \tfrac{1}{\sigma_w^2} \mathbf{H} \mathbf{Q} \mathbf{H}^{\mathsf{H}} \right), \\ \text{s.t.} \quad & \operatorname{tr}\left(\mathbf{F} \mathbf{Q} \mathbf{F}^{\mathsf{H}} \right) \leq \tilde{P}_{\mathrm{L}}, \\ & \operatorname{tr}\left(\mathbf{Q} \right) \leq \tilde{P}_{\mathrm{Tx}}, \end{aligned} \tag{4.7}$$

with $\tilde{P}_{\mathrm{Tx}}$ the transmit power constraint and $\tilde{P}_{\mathrm{L}}$ the leakage power constraint. This optimization problem was similarly but independently proposed in [62], however without providing a closed form solution and analysis of the resulting relations.

For any closed convex constraint set, the rate optimal point lies on the boundary of the set, as otherwise we could always increase the leakage and/or the transmit power and achieve a higher rate until we reach the boundary (as discussed in the previous subsection). For the special case of a rectangular constraint set $P_{\mathrm{L}} \leq \tilde{P}_{\mathrm{L}}$, $P_{\mathrm{Tx}} \leq \tilde{P}_{\mathrm{Tx}}$ as given in the optimization problem (4.7), we can distinguish three situations. If $\tilde{P}_{\mathrm{L}}$ is higher than the resulting leakage power $\hat{P}_{\mathrm{L}}$ of the rate maximization under a $\tilde{P}_{\mathrm{Tx}}$ constraint (egoistic case), the transmit power is the dominating constraint. Hence, the rate optimal point lies on the solid red curve in Fig. 4.2. If $\tilde{P}_{\mathrm{Tx}}$ is higher than the resulting transmit power $\hat{P}_{\mathrm{Tx}}$ of the rate maximization under a $\tilde{P}_{\mathrm{L}}$ constraint (altruistic case), the leakage power is the dominating constraint. In this case, the rate optimal point lies on the solid green curve. For all other cases, both constraints are restraining and the rate optimal point can be found on the intersection of the two constraints, marked with the blue cross in Fig. 4.2. For a reasonable choice of the constraints no resources are wasted. If a link generates a very low leakage level only the transmit power constraint would be active. At the same time the network is protected from very high leakage levels due to the leakage power constraint. The solutions for the respective cases can be found as follows.

The egoistic case In this case, we consider only a transmit power constraint $\tilde{P}_{\mathrm{Tx}}$. That is, the precoding is designed in an egoistic way without any considerations about the interference produced for the rest of the system. Hence, the optimization problem reduces to

$$\max_{\mathbf{Q}} \log_2 \det\left(\mathbf{I} + \frac{1}{\sigma_w^2}\mathbf{H}\mathbf{Q}\mathbf{H}^{\mathsf{H}}\right) \text{ s.t. } \operatorname{tr}(\mathbf{Q}) \leq \tilde{P}_{\mathrm{Tx}}. \tag{4.8}$$

This corresponds to the standard point-to-point MIMO problem. By diagonalizing the problem using the singular value decomposition of $\mathbf{H}$, $\mathbf{H} = \mathbf{U_H}\mathbf{S_H}\mathbf{V}_{\mathbf{H}}^{\mathsf{H}}$, and precoding with $\mathbf{W} = \bar{\mathbf{V}}_{\mathbf{H}}$, the optimization problem can be brought into the equivalent form

$$\max_{\mathbf{\Lambda}_{\mathrm{s}}} \log_2 \det\left(\mathbf{I} + \frac{1}{\sigma_w^2}\bar{\mathbf{S}}_{\mathbf{H}}\mathbf{\Lambda}_{\mathrm{s}}\bar{\mathbf{S}}_{\mathbf{H}}^{\mathsf{H}}\right) \text{ s.t. } \operatorname{tr}(\mathbf{\Lambda}_{\mathbf{s}}) \leq \tilde{P}_{\mathrm{Tx}}, \tag{4.9}$$

where $\bar{\mathbf{S}}_{\mathbf{H}}$ denotes the part of $\mathbf{S_H}$ containing the non-zero singular values, and $\bar{\mathbf{V}}_{\mathbf{H}}$ the corresponding columns of $\mathbf{V_H}$. This problem is maximized by a diagonal $\mathbf{\Lambda}_{\mathrm{s}}$ with entries found by waterfilling over the diagonal elements of $\bar{\mathbf{S}}_{\mathbf{H}}^{\mathsf{H}}\bar{\mathbf{S}}_{\mathbf{H}}$ [5]. This optimization leads to the maximal achievable rate at the mobile of interest under the transmit power constraint, while disregarding the leakage power.

The altruistic case In the altruistic case, we only have a leakage power constraint and the optimization problem reduces to

$$\max_{\mathbf{Q}} \log_2 \det\left(\mathbf{I} + \frac{1}{\sigma_w^2}\mathbf{H}\mathbf{Q}\mathbf{H}^{\mathsf{H}}\right) \text{ s.t. } \operatorname{tr}\left(\mathbf{F}\mathbf{Q}\mathbf{F}^{\mathsf{H}}\right) \leq \tilde{P}_{\mathrm{L}}. \tag{4.10}$$

It is important to note that we need to have $N_{\mathrm{VM}} \geq N_{\mathrm{BS}}$ for this optimization problem. Otherwise the leakage can be nulled and an infinite rate can be achieved (as there is no transmit power constraint). In order to solve problem (4.10), we can not resort to the singular value decomposition, as the leakage power constraint would not simplify as desired. Instead, following [18], we can resort to the generalized eigenvalue decomposition (GEVD) to diagonalize the system, $\mathbf{H}^{\mathsf{H}}\mathbf{H}\mathbf{Z} = \mathbf{F}^{\mathsf{H}}\mathbf{F}\mathbf{Z}\mathbf{D}$, where $\mathbf{D}$ contains the generalized eigenvalues of the matrix pair $\left(\mathbf{H}^{\mathsf{H}}\mathbf{H}, \mathbf{F}^{\mathsf{H}}\mathbf{F}\right)$, and $\mathbf{Z}$ the corresponding generalized eigenvectors. In order to do so, we first take a step back for reasons of clarity, and consider a signal vector $\check{\mathbf{s}} \in \mathbb{C}^{N_{\mathrm{BS}}}$ with $\check{\mathbf{\Lambda}}_{\mathrm{s}} = \mathsf{E}\left[\check{\mathbf{s}}\check{\mathbf{s}}^{\mathsf{H}}\right]$ instead of $\mathbf{s}$ in Fig. 4.1. This requires a precoding matrix $\check{\mathbf{W}} \in \mathbb{C}^{N_{\mathrm{BS}} \times N_{\mathrm{BS}}}$ and results in the transmit covariance matrix $\check{\mathbf{Q}} = \check{\mathbf{W}}\check{\mathbf{\Lambda}}_{\mathrm{s}}\check{\mathbf{W}}^{\mathsf{H}}$. Maximizing the achievable rate of this system under the leakage power constraint $\tilde{P}_{\mathrm{L}}$ leads to the optimization problem

$$\max_{\check{\mathbf{Q}}} \log_2 \det \left(\mathbf{I} + \frac{1}{\sigma_w^2} \mathbf{H}\check{\mathbf{Q}}\mathbf{H}^{\mathsf{H}} \right) \text{ s.t. } \operatorname{tr}\left(\mathbf{F}\check{\mathbf{Q}}\mathbf{F}^{\mathsf{H}}\right) \leq \tilde{P}_{\mathrm{L}}. \tag{4.11}$$

If we now let $\check{\mathbf{W}} = \mathbf{Z}$, note that $\mathbf{Z}$ is invertible, and apply the properties of the generalized eigenvalue decomposition

$$\begin{aligned} \mathbf{Z}^{\mathsf{H}}\mathbf{H}^{\mathsf{H}}\mathbf{H}\mathbf{Z} &= \mathbf{D} \\ \mathbf{Z}^{\mathsf{H}}\mathbf{F}^{\mathsf{H}}\mathbf{F}\mathbf{Z} &= \mathbf{I}, \end{aligned} \tag{4.12}$$

the optimization problem can be brought into the equivalent form

$$\max_{\check{\mathbf{\Lambda}}_{\mathrm{s}}} \log_2 \det \left(\mathbf{I} + \frac{1}{\sigma_w^2} \mathbf{D}^{1/2}\check{\mathbf{\Lambda}}_{\mathrm{s}}\mathbf{D}^{1/2} \right) \text{ s.t. } \operatorname{tr}\left(\check{\mathbf{\Lambda}}_{\mathrm{s}}\right) \leq \tilde{P}_{\mathrm{L}}. \tag{4.13}$$

For $N_{\mathrm{MS}} < N_{\mathrm{BS}}$ (as in our setup) $\operatorname{rank}(\mathbf{D}) \leq N_{\mathrm{MS}}$. That is, at least $N_{\mathrm{BS}} - N_{\mathrm{MS}}$ of the diagonal elements of $\mathbf{D}$ are 0. As it does not make sense to assign transmit power to the corresponding streams in $\check{\mathbf{s}}$, we can focus on the (usually N_{MS}) non-zero generalized eigenvalues summarized in the diagonal matrix $\bar{\mathbf{D}} \in \mathbb{C}^{N_{\mathrm{MS}} \times N_{\mathrm{MS}}}$ and the signal vector $\mathbf{s} \in \mathbb{C}^{N_{\mathrm{MS}}}$. Hence, the optimization problem simplifies to

$$\max_{\mathbf{\Lambda}_{\mathrm{s}}} \log_2 \det \left(\mathbf{I} + \frac{1}{\sigma_w^2} \bar{\mathbf{D}}^{1/2}\mathbf{\Lambda}_{\mathrm{s}}\bar{\mathbf{D}}^{1/2} \right) \text{ s.t. } \operatorname{tr}\left(\mathbf{\Lambda}_{\mathrm{s}}\right) \leq \tilde{P}_{\mathrm{L}}. \tag{4.14}$$

This optimization problem is equivalent to the optimization problem in Eq. (4.10). Analogously to (4.9), it is optimized by a diagonal $\mathbf{\Lambda}_{\mathrm{s}}$ with entries found by waterfilling over the values of $\bar{\mathbf{D}}$. Hence, the solution to the altruistic case can be found by precoding with $\mathbf{W} = \bar{\mathbf{Z}}$, the generalized eigenvectors corresponding to the generalized eigenvalues in $\bar{\mathbf{D}}$, and $\mathbf{\Lambda}_{\mathrm{s}}$ as described above.

Optimization under joint constraints To capture the impact of both power constraints, we consider the weighted sum P_ϑ of the transmit power and the leakage power,

$$\begin{aligned} P_\vartheta &= \vartheta P_{\mathrm{L}} + (1-\vartheta)\, P_{\mathrm{Tx}} \\ &= \vartheta \operatorname{tr}\left(\mathbf{W}^{\mathsf{H}}\mathbf{F}^{\mathsf{H}}\mathbf{F}\mathbf{W}\mathbf{\Lambda}_{\mathrm{s}}\right) + (1-\vartheta) \operatorname{tr}\left(\mathbf{W}^{\mathsf{H}}\mathbf{W}\mathbf{\Lambda}_{\mathrm{s}}\right) \\ &= \operatorname{tr}\left(\mathbf{W}^{\mathsf{H}}\left(\vartheta\mathbf{F}^{\mathsf{H}}\mathbf{F} + (1-\vartheta)\,\mathbf{I}\right)\mathbf{W}\mathbf{\Lambda}_{\mathrm{s}}\right) \\ &= \operatorname{tr}\left(\mathbf{W}^{\mathsf{H}}\tilde{\mathbf{F}}^{\mathsf{H}}\tilde{\mathbf{F}}\mathbf{W}\mathbf{\Lambda}_{\mathrm{s}}\right) \\ &= \operatorname{tr}\left(\tilde{\mathbf{F}}\mathbf{Q}\tilde{\mathbf{F}}^{\mathsf{H}}\right). \end{aligned} \tag{4.15}$$

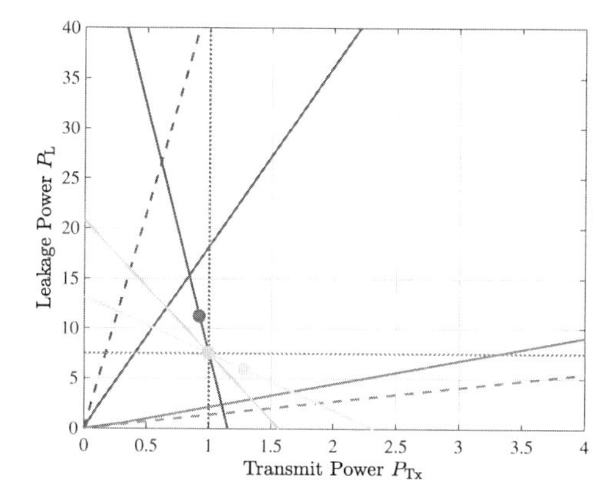

Figure 4.3.: Visualization of the line search in the transmit power - leakage power plane. The dots indicate the points with the maximal achievable rate on the respective lines.

By optimizing the rate subject to the single power constraint $\tilde{P}_\vartheta$ instead of the two individual constraints, we get the same structure in the optimization problem as in the altruistic case. Hence, we can optimize the achievable rate subject to the combined constraint $\tilde{P}_\vartheta$ in a closed form in the same way over the generalized eigenvalue decomposition of $\left(\mathbf{H}^{\mathsf{H}}\mathbf{H}, \tilde{\mathbf{F}}^{\mathsf{H}}\tilde{\mathbf{F}}\right)$, with $\tilde{\mathbf{F}}^{\mathsf{H}}\tilde{\mathbf{F}} = \left(\vartheta \mathbf{F}^{\mathsf{H}}\mathbf{F} + (1-\vartheta)\,\mathbf{I}\right)$. Note that for given ϑ and $\tilde{P}_\vartheta$, the set of points that solve (4.15) defines a line in the transmit power - leakage power plane. The optimization will then yield the point on this line which corresponds to the maximal achievable rate. As we are interested in the precoding which leads to $(\tilde{P}_{\mathrm{Tx}}, \tilde{P}_{\mathrm{L}})$ (maximum achievable rate under the two constraints at this point), we consider all lines that pass through this point and search for the one leading to the desired leakage and transmit power pair $(\tilde{P}_{\mathrm{Tx}}, \tilde{P}_{\mathrm{L}})$. This is illustrated in Fig. 4.3 where multiple such lines with their rate maxima are drawn. The set of lines through $(\tilde{P}_{\mathrm{Tx}}, \tilde{P}_{\mathrm{L}})$ is parameterized by the parameter ϑ and we choose the combined constraint as

$$\tilde{P}_\vartheta = \vartheta \tilde{P}_{\mathrm{L}} + (1-\vartheta)\,\tilde{P}_{\mathrm{Tx}}. \tag{4.16}$$

All lines with a positive slope would lead to infinite rate, as both, the transmit power and the leakage power can be increased arbitrarily. Hence, we only have to consider lines with a negative or zero slope and the optimization problem reduces to a one dimensional search over $\vartheta \in [0, 1]$. To find the optimal ϑ, we use the bisection method

Algorithm 1 Finding the rate optimal precoding with the bisection method

1: Initialization: $\vartheta_{\max} = 1$, $\vartheta_{\min} = 0$, $\hat{P}_{\mathrm{L}} = 0$, $\hat{P}_{\mathrm{Tx}} = 0$, η
2: **while** $|\, \hat{P}_{\mathrm{L}} - \tilde{P}_{\mathrm{L}} \,| > \tilde{P}_{\mathrm{L}} \cdot \eta$ or $|\, \hat{P}_{\mathrm{Tx}} - \tilde{P}_{\mathrm{Tx}} \,| > \tilde{P}_{\mathrm{Tx}} \cdot \eta$ **do**
3: $\quad \vartheta = \frac{\vartheta_{\max} - \vartheta_{\min}}{2}$
4: $\quad \tilde{P} = \vartheta \tilde{P}_{\mathrm{L}} + (1 - \vartheta)\, \tilde{P}_{\mathrm{Tx}}$
5: $\quad \tilde{\mathbf{F}}^{\mathrm{H}}\tilde{\mathbf{F}} = \vartheta \mathbf{F}^{\mathrm{H}}\mathbf{F} + (1 - \vartheta)\mathbf{I}$
6: $\quad (\mathbf{Z}, \mathbf{D}) = \mathrm{GEVD}\left(\mathbf{H}^{\mathrm{H}}\mathbf{H}, \tilde{\mathbf{F}}^{\mathrm{H}}\tilde{\mathbf{F}}\right)$
7: $\quad \max_{\mathbf{\Lambda}_{\mathrm{s}}} \log_2 \det\left(\mathbf{I} + \frac{1}{\sigma_w^2}\bar{\mathbf{D}}^{1/2}\mathbf{\Lambda}_{\mathrm{s}}\bar{\mathbf{D}}^{1/2}\right)$ s.t. $\mathrm{tr}\left(\mathbf{\Lambda}_{\mathrm{s}}\right) \leq \tilde{P}$
8: $\quad \mathbf{W} = \bar{\mathbf{Z}}$
9: $\quad \hat{P}_{\mathrm{L}} = \mathrm{tr}\left(\mathbf{F}\mathbf{W}\mathbf{\Lambda}_{\mathrm{s}}\mathbf{W}^{\mathrm{H}}\mathbf{F}^{\mathrm{H}}\right)$
10: $\quad \hat{P}_{\mathrm{Tx}} = \mathrm{tr}\left(\mathbf{W}\mathbf{\Lambda}_{\mathrm{s}}\mathbf{W}^{\mathrm{H}}\right)$
11: $\quad$ **if** $\tilde{P}_{\mathrm{L}} < \hat{P}_{\mathrm{L}}$ **then**
12: $\qquad \vartheta_{\max} = \vartheta$
13: $\quad$ **else**
14: $\qquad \vartheta_{\min} = \vartheta$
15: $\quad$ **end if**
16: **end while**

and iteratively search for the ϑ leading to $(\tilde{P}_{\mathrm{Tx}}, \tilde{P}_{\mathrm{L}})$ up to a tolerance η. The detailed procedure can be found in Algorithm 1. With this algorithm, the optimal solution is found within a few iterations, depending on η.

4.3.3 Numerical Evaluation

In the following, the performance of rate optimal precoding under a transmit power and leakage power constraint is evaluated by numerical simulations and compared to SLNR precoding [18] as a baseline of leakage based precoding where no leakage power constraint is considered.

SLNR precoding In multi-user SLNR precoding with multiple streams per user according to [18], the signal-to-leakage-plus-noise ratio is maximized for each user separately for $\mathbf{\Lambda}_{\mathrm{s}} = \mathbf{I}$, i.e. for uncorrelated streams with equal power allocation. Considering the transmit power constraint $\mathrm{tr}(\mathbf{W}^{\mathrm{H}}\mathbf{W}) = \tilde{P}_{\mathrm{Tx}}$, the SLNR for a specific user is given as

$$\mathrm{SLNR} = \frac{\mathrm{tr}\left(\mathbf{W}^{\mathrm{H}}\mathbf{H}^{\mathrm{H}}\mathbf{H}\mathbf{W}\right)}{\mathrm{tr}\left(\mathbf{W}^{\mathrm{H}}\left(\mathbf{F}^{\mathrm{H}}\mathbf{F} + N_{\mathrm{VM}} \cdot \sigma_w^2/\tilde{P}_{\mathrm{Tx}} \cdot \mathbf{I}\right)\mathbf{W}\right)}. \tag{4.17}$$

As derived in [18] this expression is maximized by the generalized eigenvectors corresponding to the N_{MS} biggest generalized eigenvalues of the generalized eigenvalue decomposition of the matrix pair $\left(\mathbf{H}^{\mathsf{H}}\mathbf{H}, \mathbf{F}^{\mathsf{H}}\mathbf{F} + N_{\mathrm{VM}} \cdot \sigma_w^2/\tilde{P}_{\mathrm{Tx}} \cdot \mathbf{I}\right)$. In contrast to the proposed approach, the leakage power level cannot be steered. The resulting operating point in the transmit power - leakage power plane will lie somewhere close to the solid greed line in Fig. 4.2 (can not be specified exactly due to the noise term in the maximization). Furthermore, there is no power loading among the streams. Hence, the rate is generally not maximized for the generated leakage power level.

Simulation setup The simulation setup is a single cell multi-user MIMO system with one base station with $N_{\mathrm{BS}} = 8$ antennas and 6 mobile stations with $N_{\mathrm{MS}} = 2$ antennas each. For each mobile station, the precoding is determined separately, considering the other mobile stations as victim mobiles. However, all mobile stations are then served simultaneously. The elements of the channel matrices $\mathbf{H}$ and $\mathbf{F}$ are assumed to be i.i.d. $\mathcal{CN}(0, 1)$. That is, the mobile stations are not subject to path loss and shadowing in this evaluation. The noise variance at the mobile stations is set to $\sigma_w^2 = 1$. Additionally to the thermal noise, the interference originating from the data streams for the other mobile stations in the cell is present at the mobile stations. As no feedback of the interference level from the mobile stations is considered, we assume for the calculation of the linear precoding that the interference is spatially white zero mean complex Gaussian distributed with variance $\sigma_{\mathrm{i}}^2 = \tilde{P}_{\mathrm{L}}/N_{\mathrm{MS}}$ at each mobile of interest antenna. Hence, $\mathbf{K}_{\mathrm{z}} = \mathrm{diag}(1 + \tilde{P}_{\mathrm{L}}/N_{\mathrm{MS}}, \ldots, 1 + \tilde{P}_{\mathrm{L}}/N_{\mathrm{MS}})$. This value would arise in practice, if all interference contributions were distributed uniformly across the victim mobile antennas. For the computation of the achievable rates, however, we consider the true interference covariance matrices. The constraints are set to $\tilde{P}_{\mathrm{Tx}} = 1$, $\tilde{P}_{\mathrm{L}} \in \{0.5, 1, \ldots, 10\}$ and $\eta = 0.001$.

Results Fig. 4.4 shows the average achievable rate per mobile station, and Fig. 4.5 the average transmit power and leakage power involved. Considering the rate maximization under joint constraints, it can be seen in Fig. 4.4 that for a low leakage power constraint $\tilde{P}_{\mathrm{L}}$ the achievable rate is inferior compared to higher leakage power constraints (up to $\tilde{P}_{\mathrm{L}} = 3$). This comes from the fact that for most channel realizations the leakage power constraint $\tilde{P}_{\mathrm{L}}$ is the dominating constraint, allowing only for a low transmit power (see Fig. 4.5). This leads to a low signal power and hence to a low achievable rate. Increasing the leakage power constraint $\tilde{P}_{\mathrm{L}}$ allows for higher transmit

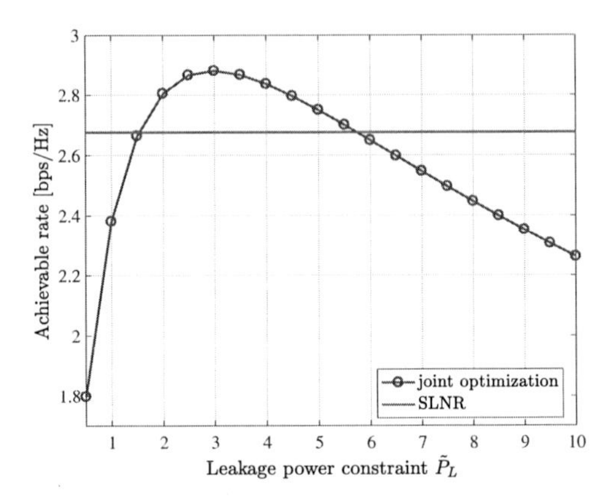

Figure 4.4.: Average achievable rate per mobile station. The SLNR precoding is independent of the leakage power constraint.

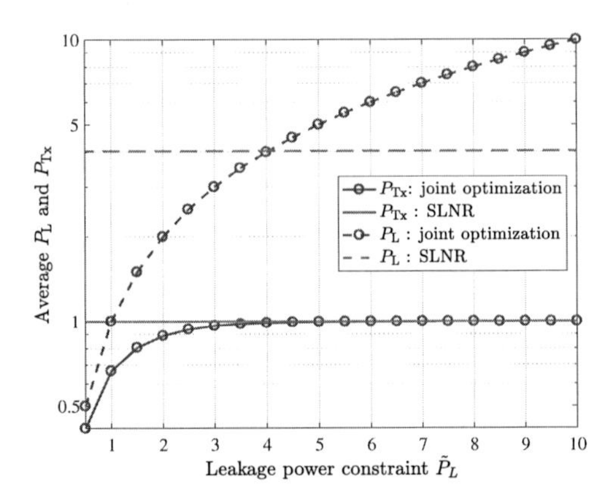

Figure 4.5.: Average transmit and leakage power per mobile station. The SLNR precoding is independent of the leakage power constraint.

power and thus higher signal power. This can compensate the degradation due to the additional interference in the network, leading to increased achievable rates. The higher the leakage power constraint is, the more channel realizations lead to the case where both constraints are restraining. However, this does not lead to a consistently increasing performance. For a certain leakage power constraint $\tilde{P}_{\mathrm{L}}$ the peak performance is reached (in this setup for $\tilde{P}_{\mathrm{L}} = 3$). For any higher leakage power constraint the disadvantage of more interference in the network can not be compensated anymore by the advantage of an increased signal power. Thus, the achievable rates are decreasing again. Compared to SLNR precoding, the performance can be clearly increased at the peak performance point using the optimization under joint constraints. As we do not consider a leakage power constraint for the SLNR precoding its results are not affected by it.

Comparing the average transmit power (Fig. 4.5) of SLNR precoding and the optimization under joint constraints at the point where both schemes achieve about the same average achievable rate (i.e. for $\tilde{P}_{\mathrm{L}} = 1.5$, see Fig. 4.4), it can be observed that significantly less transmit power is used for the optimization under joint constraints. That is, with a flexible $\tilde{P}_{\mathrm{L}}$ constraint, the network performance can be optimized and transmit power saved.

Note on complexity: Throughout our simulations in Matlab, the computation time of the proposed algorithm was about 1 to 2 orders of magnitude smaller than with a standard optimization toolbox for semidefinite programming [66]. The average number of iterations has been around 7.

Discussion As can be seen from the curves in Fig. 4.5, in most cases either the leakage power constraint is restraining or both constraints jointly. That is, the case where only the transmit power constraint is restraining, because the channel constellation causes only very low leakage, did not occur frequently due to the setup without path loss and shadowing. However, especially in these cases the proposed precoding could exploit its full potential by not wasting the energy for the leakage minimization if it is already low anyways. Hence, for a setup with strongly varying channels among the mobile stations even larger gains are expected, as very low or very high leakage levels are more likely to occur.

4.4 Leakage Based Target Rate Precoding

In multi-user MIMO systems with one or multiple transmitters and multiple receivers, the distribution of the achievable rates among the users is of crucial interest from a fairness point of view. Different beamforming strategies can be thought of. While sum rate maximization in a system maximizes the total throughput, the achievable rates can be distributed very unevenly among the different users. Strong users get very high rates and weak users very low rates or even nothing at all. Hence, it is unpractical to be applied in a real multi-user system in terms of user satisfaction. In contrast to sum rate maximization, maximizing the minimal rate (Max-Min) among all users leads to equal rates for all users, as the rate of the strong users is reduced for the benefit of the weak ones. However, this equality could lead to very low rates and low total throughput if one user has a very poor channel. Hence, this scheme is unpractical in real systems as well.

A promising alternative is target rate precoding. The precoding is designed such that a user just achieves a certain predefined target rate. This rate could be different for every user and adapted according to applications and/or user priorities. In contrast to Max-Min, the rates of the strong users are only reduced down to their target rate. If a weak user is not able to achieve its target rate, it is considered to be in an outage. However, all other users are not further affected. As will be shown later, the transmitters can thus efficiently use their resources to serve as many receivers as possible with their desired target rate, while using as little transmit power as possible.

4.4.1 Target Rate Precoding

Analogously to the rate optimal precoding in Sec. 4.3, we propose a decentralized target rate precoding. That is, each link is considered separately without knowledge about the precoding of all others. Therefore, no global optimization is necessary and only limited channel state information is required. However, the whole network is inherently coupled, as the precoding of each mobile station affects the performance of all others via the generated leakage. Hence, feedback of the interference level at the mobile stations and iterative adaption of the precoding is necessary for reliable target rate precoding. In order to control the co-channel interference, the transmit power as well as the leakage power are considered in the precoding design.

To goal of the precoding is then to minimize the transmit power as well as the leakage power while achieving a predefined target rate R_t. To capture both objective functions in one, we again consider the weighted sum power of Eq. (4.15),

$$\begin{aligned} P_\vartheta &= \vartheta P_\mathrm{L} + (1-\vartheta) P_\mathrm{Tx} \\ &= \operatorname{tr}\left(\tilde{\mathbf{F}}\mathbf{Q}\tilde{\mathbf{F}}^\mathsf{H}\right), \end{aligned} \tag{4.18}$$

with $\tilde{\mathbf{F}}^\mathsf{H}\tilde{\mathbf{F}} = \left(\vartheta \mathbf{F}^\mathsf{H}\mathbf{F} + (1-\vartheta)\mathbf{I}\right)$ and $\vartheta \in [0,1]$. Thus, for a fixed ϑ and summarizing the channel matrix and the interference plus noise covariance matrix into $\tilde{\mathbf{H}} = \mathbf{K}_\mathrm{z}{}^{-1/2}\mathbf{H}$, the optimization problem can be stated as

$$\min_{\mathbf{Q}} \operatorname{tr}\left(\tilde{\mathbf{F}}\mathbf{Q}\tilde{\mathbf{F}}^\mathsf{H}\right) \text{ s.t. } \log_2 \det\left(\mathbf{I} + \tilde{\mathbf{H}}\mathbf{Q}\tilde{\mathbf{H}}^\mathsf{H}\right) \geq R_\mathrm{t}. \tag{4.19}$$

That is, we design the precoding matrix $\mathbf{W}$ and the signal covariance matrix $\mathbf{\Lambda}_\mathrm{s}$ such that the weighted sum power is minimized under the target rate constraint.

This problem is solved analogously to the maximization of the achievable rate under a leakage power constraint in Eq. (4.10). The generalized eigenvalue decomposition of the matrix tuple $\left(\tilde{\mathbf{H}}^\mathsf{H}\tilde{\mathbf{H}}, \tilde{\mathbf{F}}^\mathsf{H}\tilde{\mathbf{F}}\right)$ provides a diagonal matrix $\mathbf{D}$ containing the generalized eigenvalues of the tuple and a matrix $\mathbf{Z}$ containing the corresponding generalized eigenvectors as columns, such that

$$\begin{aligned} \mathbf{Z}^\mathsf{H}\tilde{\mathbf{H}}^\mathsf{H}\tilde{\mathbf{H}}\mathbf{Z} &= \mathbf{D} \\ \mathbf{Z}^\mathsf{H}\tilde{\mathbf{F}}^\mathsf{H}\tilde{\mathbf{F}}\mathbf{Z} &= \mathbf{I}. \end{aligned} \tag{4.20}$$

Hence, by substituting $\mathbf{W} = \bar{\mathbf{Z}}$, where $\bar{\mathbf{Z}}$ corresponds to the nonzero generalized eigenvalues $\bar{\mathbf{D}}$ in $\mathbf{D}$, we can find the equivalent optimization problem

$$\min_{\mathbf{\Lambda}_\mathrm{s}} \operatorname{tr}\left(\mathbf{\Lambda}_\mathrm{s}\right) \text{ s.t. } \log_2 \det\left(\mathbf{I} + \bar{\mathbf{D}}^{\frac{1}{2}}\mathbf{\Lambda}_\mathrm{s}\bar{\mathbf{D}}^{\frac{1}{2}}\right) \geq R_\mathrm{t}. \tag{4.21}$$

This problem is optimized by a diagonal $\mathbf{\Lambda}_\mathrm{s}$. The proof thereof works analogously to the proof of the point to point MIMO capacity in [5]. Due to the diagonal structure of $\mathbf{\Lambda}_\mathrm{s}$, its elements can be found by splitting up the system into parallel channels and applying the method of Lagrange multipliers:

$$\lambda_{i,i} = \left(\mu - \frac{1}{\delta_i}\right)^+ \tag{4.22}$$

with the Lagrange multiplier μ such that

$$\sum_{i=1}^{N_{\mathrm{MS}}} \log_2 \left(1 + \lambda_{i,i}\delta_i\right) = R_{\mathrm{t}}. \tag{4.23}$$

Here, $\lambda_{i,i}$ are the entries of the diagonal matrix $\mathbf{\Lambda}_{\mathrm{s}}$, δ_i the generalized eigenvalues, and $(x)^+ = \max(0, x)$.

To get a better understanding of this optimization problem, we have a look at the transmit power - leakage power plane in Fig. 4.6. It shows an example outcome of the relation between the leakage power and the transmit power. The solid red and green curves show the resulting leakage power respectively transmit power if the rate at the mobile of interest is maximized under a transmit respectively leakage power constraint as introduced in Sec. 4.3.2. For a fixed ϑ, the optimization minimizes the weighted sum power subject to the target rate. The choice of ϑ allows to control the relation of the transmit power and leakage power minimization and different points in the transmit power - leakage power plane can be achieved. For $\vartheta = 0$, the transmit power is minimized and hence, the resulting transmit and leakage power lie on the red curve. For $\vartheta = 1$, the leakage power is minimized and the resulting transmit and leakage power lie on the green curve. For any ϑ in between, the resulting $(P_{\mathrm{Tx}}, P_{\mathrm{L}})$ tuples lie somewhere in between and describe a curve, further called the *target rate curve*. That is, the target rate R_{t} can be achieved with any transmit and leakage power pair on this curve. It can be observed that the lower the transmit power, the higher the generated leakage power and vice versa. However, the curve is rather flat for high transmit power. That is, in this area a further reduction of the leakage power comes at a high price of additional transmit power. This relation is shown for different target rates ($R_{\mathrm{t}} \in \{2, 3, 4, 5\}$ bps/Hz) in Fig. 4.6. Obviously, a higher target rate leads to higher transmit and leakage power for a fixed interference and noise level. Hence, the target rate curve is shifted up and to the right for increasing target rate R_{t}.

4.4.2 The Transmit Power - Leakage Power Trade-Off

For a mobile station in a single-user MIMO setup as shown in Fig. 4.1 with fixed $\mathbf{H}$ and $\mathbf{K}_{\mathrm{z}}$, a given target rate can be achieved by choosing any point on the target rate curve. However, in practical systems the transmit power is constraint, thus not all points on the curve might be feasible or non at all. Furthermore, as the whole network is inherently coupled via the generated leakage, the precoding of each mobile station

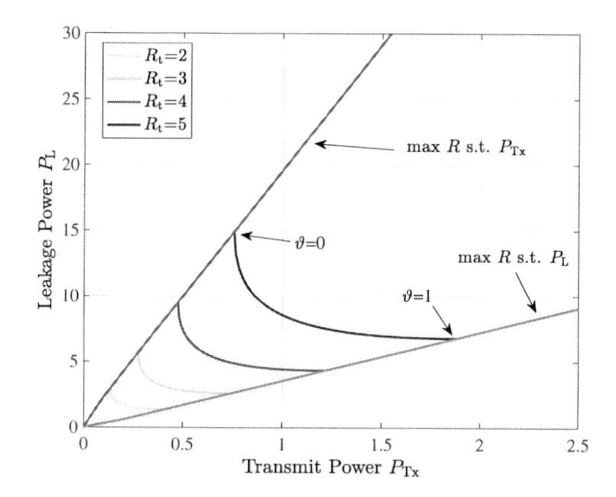

Figure 4.6.: Target rate curves in the transmit power - leakage power plane.

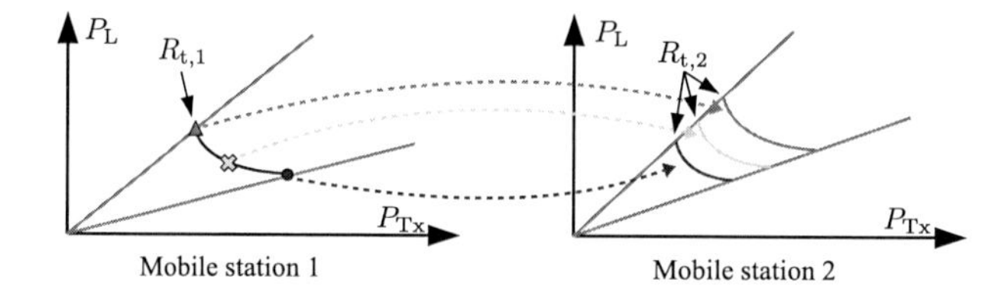

Figure 4.7.: Illustration of the inherent coupling of the mobile stations in the network.

affects the performance of all other mobile stations. This is illustrated in Fig. 4.7. Hence, depending on the resulting interference plus noise level and the transmit power constraint, a mobile station could be unable to achieve its target rate, leading to an outage.

To optimize the performance of the network in terms of outage minimization and energy efficiency, it is important to find a good trade-off between the used transmit power and the generated leakage power, and to adapt this trade-off to the network properties such as user density, regulations etc. In different words: we have to reduce the leakage power as much as necessary, but as little as possible.

As the interference level at a mobile station changes with every update of the precoding for the other mobile stations, iterative adaption of the precoding is necessary for a reasonable performance. To this end, feedback of the interference plus noise covariance matrix $\mathbf{K}_\mathbf{z}$ is considered. After each feedback, the precoding is updated, resulting in more or less transmit power and generating more or less leakage power. Therefore, convergence of the precoding is not guaranteed. However, it is shown in Sec. 4.4.3 that the system stabilizes already after a few iterations if a reasonable strategy is chosen.

In the following, we consider three different strategies for the transmit power - leakage power trade-off:

1. In strategy 1 (S_1), we always choose the $(P_{\mathrm{Tx}}, P_{\mathrm{L}})$ tuple with minimal transmit power. That is, we set $\vartheta = 0$ and obtain the solution on the red curve in Fig. 4.6. If that violates the transmit power constraint $\tilde{P}_{\mathrm{Tx}}$, the obtained transmit signal is scaled to the maximal allowed transmit power.
2. In strategy 2 (S_2), we always choose the $(P_{\mathrm{Tx}}, P_{\mathrm{L}})$ tuple with minimal leakage power. That is, we always choose $\vartheta = 1$, and obtain the solution on the green curve in Fig. 4.6. Again, if this solution violates the transmit power constraint $\tilde{P}_{\mathrm{Tx}}$, the obtained transmit signal is scaled to the maximal allowed transmit power.
3. In strategy 3 (S_3), we try to find a good compromise between S_1 and S_2. The idea is to reduce the transmit power as long as we are in the flat regime of the target rate curve, i.e. as long as the reduction of the transmit power comes at a low price of increased leakage power. Therefore, we always compute the precoding for $\vartheta = 1$, i.e. the solution on the green curve in Fig. 4.6, and then reduce the used transmit power by a certain percentage by walking back on the curve of constant rate. If it is not possible to reduce the transmit power by that amount or if this point violates the transmit power constraint, we take the obtained solution for $\vartheta = 1$ and scale it to the allowed transmit power. This strategy allows to reduce the transmit power compared to S_2 and to reach the target rate for some mobile stations where S_2 would not succeed. However, it also generates more leakage.

S_1 and S_2 are simple to implement. We just calculate the solutions for $\vartheta = 0$, respectively $\vartheta = 1$, check whether the transmit power constraint is fulfilled and scale it if necessary. The solution of S_3 can be found with the bisection method. That is, we first compute the precoding for $\vartheta = 1$ and then iteratively search for the ϑ leading to the desired transmit power.

4.4.3 Numerical Evaluations

To evaluate the performance of the proposed target rate precoding, numerical simulations have been performed in a cellular network structure as shown in Fig. 4.8,

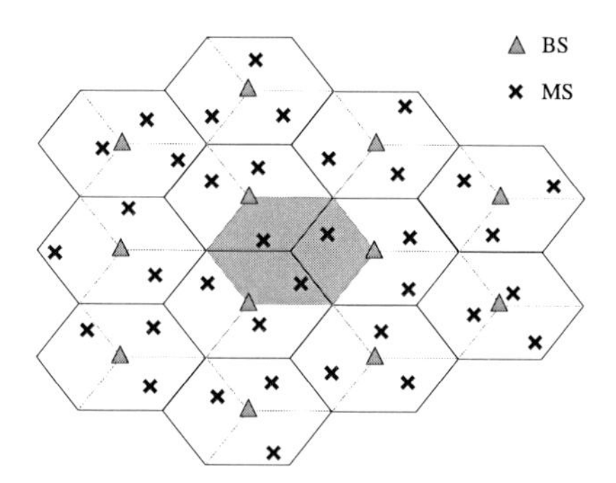

Figure 4.8.: Cellular network layout.

which was motivated in [67]. The 12 hexagonal cells with 3 sectors each are regularly arranged with a base station in the center of each cell consisting of 3 (one per sector) independent antenna arrays with $N_{\mathrm{BS}} = 8$ elements. Each of these antennas has a directional pattern (120 degree beam width) according to [68] and is oriented towards the center of the respective sector. The pattern is described in detail in App. A.1. In every sector, one mobile station with $N_{\mathrm{MS}} = 2$ omnidirectional antennas is randomly placed. These mobile stations are then served by 3 cooperating base stations applying joint beam forming. The reuse factor in the network is 1, i.e. all mobile stations are served in the same frequency band and time slot. A backhaul with infinite capacity is assumed between the cooperating base stations to share the channel state information and the transmit symbols. The cooperating base stations are grouped into fixed clusters, always consisting of 3 neighboring sectors of 3 different cells as highlighted in Fig. 4.8. Each mobile station is served with at most $\tilde{P}_{\mathrm{Tx}} = 10$ W over a bandwidth of 100 MHz around the carrier frequency $f_c = 2.6$ GHz. The distribution of this transmit power among the cooperating base stations can be arbitrary. The noise variance is assumed to be $\sigma_w^2 = 5 \cdot 10^{-12}$ W. The channels are considered frequency flat fading and are modeled by Rayleigh-fading with path loss and shadowing according to the WINNER II channel model, scenario C2 [69]. This channel model is summarized in App. A.2.

In the following, the performance of target rate precoding is investigated and the three different strategies are compared. For reasons of simplicity, the target rate is assumed to be the same for all mobile stations. In S_3, the reduction of the transmit power has been set to 40 percent for all mobile stations ($\mathrm{S}_3(0.4)$). The precoding is adapted iteratively. In each iteration, the precoding is computed for all mobile stations, then the resulting $\mathbf{K_z}$ are determined and fed back. The initial $\mathbf{K_z}$ is assumed to be $\mathbf{I} \cdot 10^{-10}$ W for all mobile stations (empirical value).

Fig. 4.9(a) and Fig. 4.9(b) show the inherent coupling of the network. S_1 has been applied to all mobile stations for $R_\mathrm{t} = 5$ bps/Hz. After the system stabilized, one mobile station in the central cluster (orange in Fig. 4.8) achieving the target rate was chosen and the ϑ adapted such that different points on the target rate curve $(P_{\mathrm{L},1}, \cdots, P_{\mathrm{L},8})$ were achieved (see Fig. 4.9(a)). That is, the generated leakage power has been decreased and the transmit power increased, while the achievable rate of this mobile station stayed constant. The effect of these changes on a second mobile station in this cluster can be observed in Fig. 4.9(b). The more the leakage at the first mobile station is decreased, the higher is the achievable rate at the second mobile station, however at the price of additional transmit power at the first mobile station. Furthermore, it can be observed that at $P_{\mathrm{L},5}$ the achievable rate starts to saturate. That is, by further decreasing the leakage power and increasing the transmit power at the first mobile station, only little can be gained at the second mobile station for a high price of additional transmit power (more than 50 percent higher). This shows how crucial it is to find a good trade-off between used transmit power and generated leakage power for the network performance.

Fig. 4.10 shows the empirical cumulative distribution functions (CDFs) of the convergence behavior for a mobile station in the central cluster if strategy 3 ($\mathrm{S}_3(0.4)$) is applied. Expectedly, the performance of the initial precoding without any knowledge about the interference plus noise level does not lead to satisfying results. However, already after two iterations of feedback, very reasonable results are achieved. After 9 iterations the system stabilizes with low outage probability (below 1 percent). The jump from 3 bps/Hz to 4 bps/Hz in the final empirical CDF is the effect of walking back on the target rate curve. By walking back, many mobile stations which would not have achieved the target rate before (as they would have violated the transmit power constraint) are now able to achieve it. The deviation from the target rate at the top of the empirical CDF comes from the fact that although the system becomes very stable, still some $\mathbf{K_z}$ change with the precoding update, and thus some mobile stations *overshoot* the target rate.

Fig. 4.11 shows the empirical CDFs for various target rates after 9 iterations of feedback. It can be observed that for all target rates the system converged. As to be expected, the outage probability continuously increases for increasing target rate.

Fig. 4.12 and Fig. 4.13 show the outage probability and the average transmit

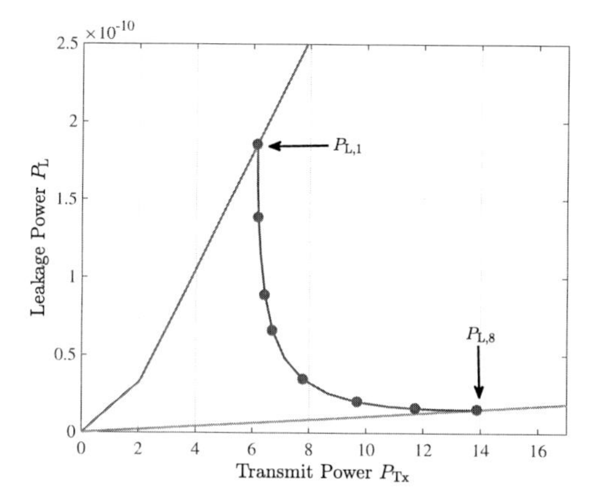

(a) Transmit power - leakage power plane of mobile station 1 for $R_t = 5$ bps/Hz.

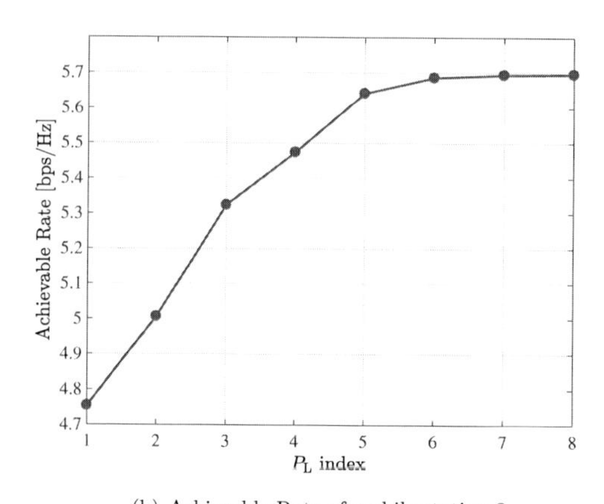

(b) Achievable Rate of mobile station 2.

Figure 4.9.: Mutual coupling of two mobile stations in the orange cluster of the network (see Fig. 4.8).

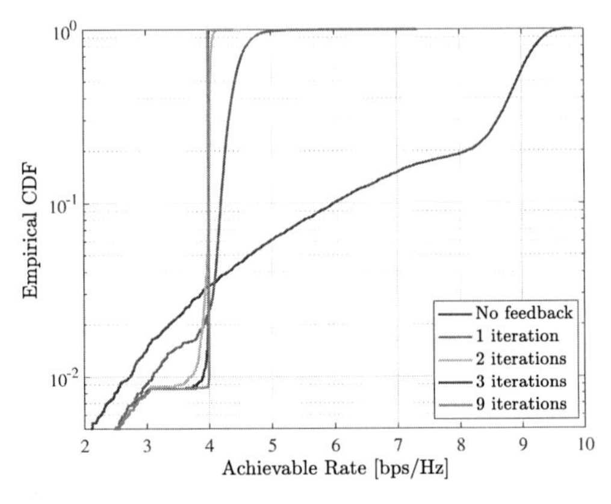

Figure 4.10.: Convergence behaviour of the achievable rates for $S_3(0.4)$ $R_\mathrm{t} = 4$ bps/Hz.

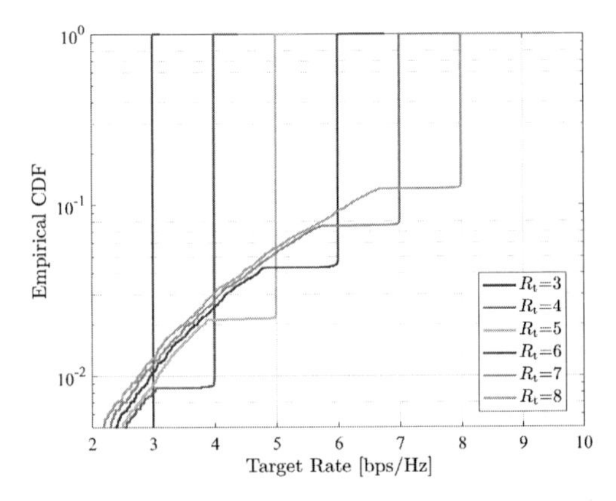

Figure 4.11.: CDFs of achievable rates for $S_3(0.4)$ and various R_t.

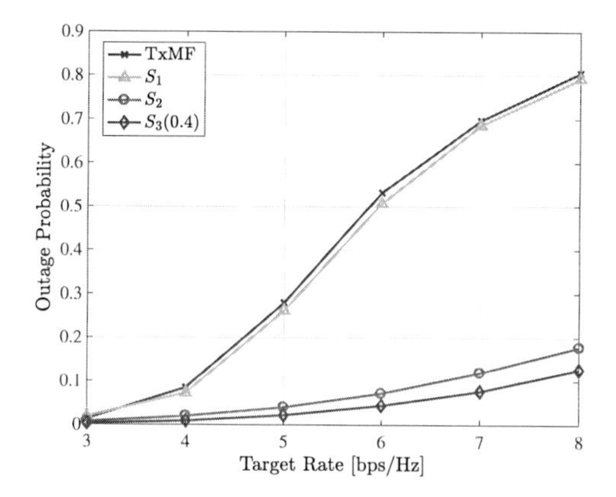

Figure 4.12.: Outage probabilities.

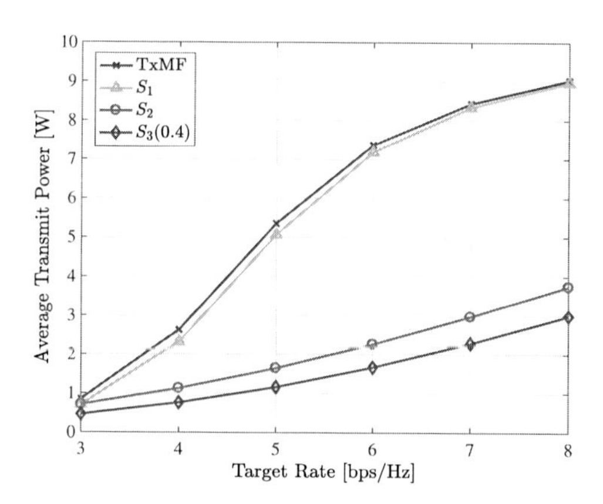

Figure 4.13.: Average transmit power.

power of all three strategies. Additionally, the performance of transmit matched filter (TxMF) precoding ($\mathbf{W} = \mathbf{H}^{\mathsf{H}}$) with power control is included for comparison. The transmit matched filter is scaled such that R_{t} is just achieved. If that is not feasible, it is scaled to $\tilde{P}_{\mathrm{Tx}}$. It can be observed that S_1 and the transmit matched filter lead to high outage probabilities and transmit power as they introduce strong leakage into the network. The increased interference level requires higher transmit power to achieve the target rate, which again generates more leakage, and so on, until the system stabilizes on a high transmit and leakage power level. Hence, although S_1 uses the least transmit power for a single-user MIMO system, it uses the most in a coupled multi-user system. S_2 and S_3 are close to each other. However, S_3 achieves up to 5 percent lower outage probability while using up to 30 percent less transmit power on average. By walking back on the target rate curve, more mobile stations achieve R_{t} although more leakage is generated. As long as we stay in the *flat* area of the curve, the benefit of achieving the target rate outweighs the loss of additional leakage power. Therefore, it is crucial to choose a reasonable value of transmit power reduction. The best option would be to adapt this value to each mobile station and its channel realization. However, this has not been considered in the scope of this thesis.

4.5 Conclusions

In this chapter, we investigated leakage based precoding as a promising multi-user MIMO precoding scheme and provided a thorough analysis of the relation of the transmit power and the resulting leakage power. Based on this relation, we proposed rate optimal precoding under a joint transmit power and leakage power constraint and derived a quasi closed-form, low complexity solution to it. Due to the leakage power constraint, the co-channel interference can be balanced and the network performance optimized. Furthermore, we also proposed a decentralized target rate precoding based on the relation between the transmit power and generated leakage power. Analogously to the rate optimal precoding, a quasi closed-form solution was derived. This precoding allows to optimize the network in terms of outage probability for specific target rates of the mobile stations and leads to promising results at low complexity.

Both schemes are well suited to be applied in cellular networks with coordinate multi point transmission. The precoding can be flexibly distributed over several base stations and also allows overlapping clusters. Equivalently, they can also be efficiently applied in any other multi-user MIMO setup.

5 Leakage Based Beam Shaping

In this chapter, we focus on transmit cooperation for range extension in mobile ad hoc networks (MANETs) by exploiting the array gain of multi-antenna systems.

To overcome large distances in MANETs a multi hop forwarding scheme is usually applied. That is, the message is transmitted from the source over multiple relays to the destination. This requires a sophisticated routing and introduces a long delay (depending on the number of necessary hops) and thus blocks the network resources for a long time. Furthermore, the multi hop transmission might fail due to node failure and the lack of alternative routes. In order to avoid these drawbacks multiple nodes could cooperate, form a virtual antenna array and optimize their radiation pattern, similar to phased arrays [9]. That is, they first exchange their transmit data and then jointly transmit the signal. This way, large gains in the radiated power can be achieved in the desired direction, leading to a larger communication range. Furthermore, by appropriate amplitude and phase excitation, the radiated power in undesired directions (further called leakage) can be minimized. This is useful to decrease the interference into the network, and - in the context of military MANETs - can reduce the probability of being detected and localized by hostile units. To optimize the radiation pattern only the relative path length differences of the transmitting nodes in the considered directions have to be available (e.g., from position information) and the nodes have to be synchronized (in time, frequency and phase). No channel state information needs to be available at the transmitters. However, due to the random node placement and the potentially large separation between the nodes in a military MANET, the radiation pattern becomes irregular and spiky, with sidelobes that are hard to control.

In this context, we propose leakage based beam shaping, a low complexity pattern synthesis approach based on the maximization of the signal-to-leakage ratio [70]. It minimizes the leakage power (signal power in undesired directions) while maintaining coherent addition in the desired direction. With a single iteration in the pattern design the transmission range can be extended while the interference into the network is minimized and the probability of being detected is reduced. We discuss the trade-off between the signal enhancement and the leakage suppression and thoroughly evaluate the proposed scheme. We furthermore address various practical considerations of leakage based beam shaping in military networks.

Analogously to transmit arrays, the same principles can be applied for receive arrays in order to suppress jammers while amplifying the received signal from the desired direction. However, we focus on the transmit cooperation in the following.

This chapter is mainly based on our work published in [34] and expanded with additional results and discussions. The remainder of this chapter is structured as follows. In Sec. 5.1 related work in the field of pattern design is reviewed. In Sec. 5.2 the scenario and system setup is discussed and in Sec. 5.3 leakage based beam shaping is introduced. In Sec. 5.4 the performance of leakage based beam shaping is evaluated and in Sec. 5.5 practical considerations and implementation issues are discussed. Sec. 5.6 finally concludes this chapter.

5.1 Related Work

Various methods for the pattern synthesis have been introduced for uniform linear arrays (ULA) and uniform rectangular arrays (URA), such as the Fourier transform method, the Dolph-Chebyshev synthesis and many others [9,71]. However, different to uniform linear arrays and uniform rectangular arrays, the nodes (and thus the antennas) are not regularly arranged in MANETs. Therefore, the above mentioned pattern synthesis algorithms can not be used.

A promising alternative are adaptive array algorithms, which allow to optimize the radiation pattern for arbitrary array geometries [72]. In an adaptive array algorithm, the pattern is iteratively optimized with respect to a desired pattern. In [73], e.g., the pattern is designed such that the main lobe is steered in a desired direction while the sidelobes meet a specific criterion. To this end, the algorithm iteratively introduces a large number of artificial interferers in the direction of the sidelobes and maximizes the signal-to-interference-plus-noise ratio until the error between the desired pattern and the synthesized pattern is minimized.

An overview of adaptive array theory can be found in [72]. Various approaches have been proposed in literature such as [73–75], leading to promising results. However, a

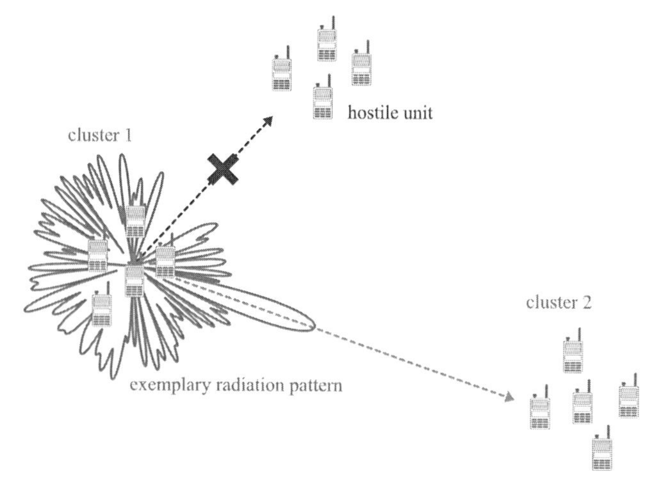

Figure 5.1.: Considered scenario for beam shaping: Two spatially separated units communicate over a large distance by forming a virtual antenna array and applying beam shaping such that the leakage into hostile directions is suppressed.

large number of iterations is necessary for the pattern synthesis, requiring high computational power. Thus, they are not well suited for the setup at hand, where frequent updates of the radiation pattern may be required due to the node mobility.

5.2 Scenario and System Setup

To investigate the potential of cooperative transmission for range extension, we consider a setup as shown in Fig. 5.1. A node in cluster 1 wants to communicate over a large distance to a node in cluster 2. Using the concepts of phased arrays, the pattern shall be designed such that the range is extended, the signal power in the direction of the hostile units is suppressed, and the interference into the network is minimized. To achieve the predicted gains, the transmit cluster should see the destination under a low angular spread in the desired direction, such that focusing the energy in this direction is reasonable. Ideally, the transmit array would be on an elevated position (e.g., a hill), with line of sight to the destination.

In the following, we consider N nodes randomly distributed in a square of side length a centered around the origin, with $x_i, y_i \in [-\frac{a}{2}, \frac{a}{2}]$ the x- and y-coordinate of node i,

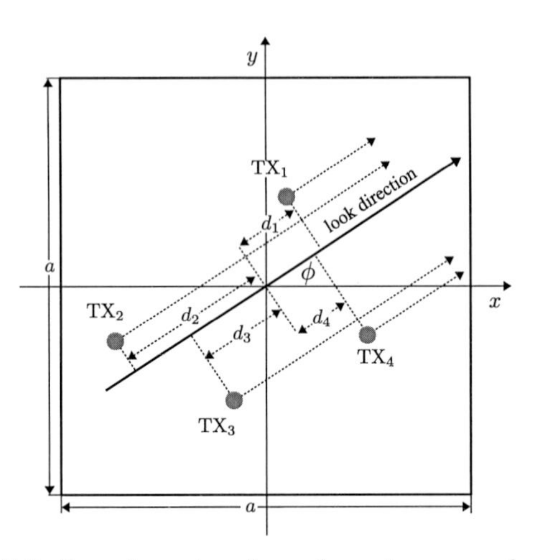

Figure 5.2.: Exemplary setup of a random antenna array for $N = 4$.

$i \in \{1, \ldots, N\}$, as shown in Fig. 5.2. All nodes are assumed to be equipped with an identical, single, omnidirectional antenna. The minimum pairwise separation between the nodes is considered to be $\lambda/2$, with λ the wavelength. That is, the antennas are uncoupled. Hence, the resulting electro-magnetic field can be considered as the superposition of the individual fields of all antenna elements. The far field array factor in direction ϕ is then given as ([71], Chap. 6)

$$\begin{aligned} Y(\phi) &= \sum_{i=1}^{N} A_i e^{-j\xi_i} e^{-j\frac{2\pi}{\lambda} d_i(\phi)} \\ &= \bar{\mathbf{v}}\mathbf{w}, \end{aligned} \tag{5.1}$$

where $d_i(\phi) = (\cos\phi \cdot x_i + \sin\phi \cdot y_i)$ denotes the path length difference of node i with respect to the origin, A_i the amplitude excitation and ξ_i the phase excitation of node i, summarized into the weight vector

$$\mathbf{w} = \left[A_1 e^{-j\xi_1}, \ldots, A_N e^{-j\xi_N}\right]^{\mathsf{T}}, \tag{5.2}$$

and the steering vector

$$\bar{\mathbf{v}} = \left[e^{-j\frac{2\pi}{\lambda} d_1(\phi)}, \ldots, e^{-j\frac{2\pi}{\lambda} d_N(\phi)}\right]. \tag{5.3}$$

The radiation pattern of the antenna array is given as the radiation pattern of a single antenna multiplied with the array factor ([71], Chap. 6). Hence, as we consider omnidirectional antennas, the gain in radiated power compared to an omnidirectional antenna transmitting at unit power is given as

$$D(\phi) = |Y(\phi)|^2. \tag{5.4}$$

The only parameters which can be influenced are the amplitude and phase excitations (assuming that the locations are fixed). That is, in order to design a certain radiation pattern, the weight vector $\mathbf{w}$ has to be optimized as discussed in the following section.

5.3 Pattern Synthesis

Motivated by the artificial interferers in [73], we consider M different look directions ϕ_m uniformly spaced on $[0, 2\pi)$, i.e. $\phi_m = (m-1)\cdot\frac{2\pi}{M}$, $m \in \{1, \ldots, M\}$. The corresponding steering vectors, given as

$$\bar{\mathbf{v}}_m = \left[e^{-j\frac{2\pi}{\lambda}d_1(\phi_m)}, \ldots, e^{-j\frac{2\pi}{\lambda}d_N(\phi_m)}\right], \tag{5.5}$$

are stacked into the matrix

$$\mathbf{V} = \begin{bmatrix} \bar{\mathbf{v}}_1 \\ \vdots \\ \bar{\mathbf{v}}_M \end{bmatrix}. \tag{5.6}$$

Each look direction is then either assigned to the desired directions, in which the radiated power shall be enhanced, or to the undesired directions, in which the leakage shall be suppressed. Furthermore, they are individually weighted in order to emphasize certain directions for the range extension or the leakage suppression. The channel in the desired directions can thus be written as

$$\mathbf{H} = \mathbf{C}_{\mathrm{H}} \cdot \mathbf{V}, \tag{5.7}$$

where $\mathbf{C}_{\mathrm{H}}$ is a diagonal matrix containing the weights of the desired directions. That is, all elements on the diagonal corresponding to undesired directions are 0. Hence, the corresponding rows in $\mathbf{H}$ are set to the zero vector. Analogously, the channel in the

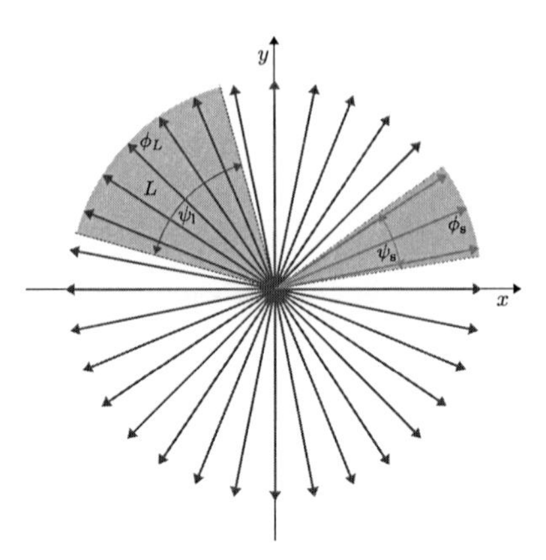

Figure 5.3.: Exemplary setup for leakage based beam shaping: M look directions indicated by the arrows, one desired direction ϕ_s with desired beam width ψ_s in green, and the undesired directions in red, with a special emphasis on the sector L of width ψ_l centered around ϕ_L.

undesired directions can be written as

$$\mathbf{F} = \mathbf{C}_{\mathrm{F}} \cdot \mathbf{V}, \tag{5.8}$$

with $\mathbf{C}_{\mathrm{F}}$ the diagonal weighting matrix for the undesired directions. Arbitrary setups and weightings can be considered. An exemplary setup matching our scenario of interest is shown in Fig. 5.3. The transmission range in the direction of ϕ_s shall be extended, while the leakage to hostile units supposed to be in the direction ϕ_L shall be strongly suppressed. To this end, we consider the desired beam width ψ_s centered around ϕ_s. All look directions in this sector are assigned to the desired channel. All other look directions are assigned to the undesired directions. For the emphasis of the leakage suppression in the direction of the hostile units, a sector L of width ψ_l centered around ϕ_L is introduced, in which the corresponding steering vectors ϕ_m are strongly weighted.

Note that the channels $\mathbf{H}$ and $\mathbf{F}$ do not contain any information about the specific channel characteristics like fading or path loss to the destination or the hostile units, as we do not consider to have this information available. It only includes the radiation pattern in the far field. Hence, there is no point in introducing any signal or leakage power constraints for the optimization of the weight vector $\mathbf{w}$.

Instead, we rather optimize the weight vector $\mathbf{w}$ such that the signal-to-leakage-power ratio as introduced in [70] is maximized. Considering transmit symbols s with $\mathsf{E}[|s|^2] = 1$, the weighted signal power (weighted power dissipated in the desired directions) can be stated as

$$P_\mathrm{s} = \mathbf{w}^\mathsf{H}\mathbf{H}^\mathsf{H}\mathbf{H}\mathbf{w}, \tag{5.9}$$

and the total weighted leakage power (weighted power dissipated in undesired directions) as

$$P_\mathrm{L} = \mathbf{w}^\mathsf{H}\mathbf{F}^\mathsf{H}\mathbf{F}\mathbf{w}. \tag{5.10}$$

The optimization problem can then be stated as

$$\hat{\mathbf{w}} = \arg\max_{\mathbf{w}} \frac{\mathbf{w}^\mathsf{H}\mathbf{H}^\mathsf{H}\mathbf{H}\mathbf{w}}{\mathbf{w}^\mathsf{H}\mathbf{F}^\mathsf{H}\mathbf{F}\mathbf{w}}. \tag{5.11}$$

The optimal $\hat{\mathbf{w}}$ can be found as a scaled version of the eigenvector corresponding to the largest eigenvalue of the generalized eigenvalue decomposition of $\mathbf{H}^\mathsf{H}\mathbf{H}$ and $\mathbf{F}^\mathsf{H}\mathbf{F}$ as shown in [70]. Hence, no iterative optimization is necessary and only low computational complexity is required. As the optimization problem is independent of the transmit power, $\hat{\mathbf{w}}$ can finally be scaled to any desired transmit power, still maximizing the signal-to-leakage-power ratio.

In Fig. 5.4 various resulting patterns are shown for a setup as sketched Fig. 5.3, considering $N \in \{5, 15, 25\}$ cooperating nodes, $M = 3600$ look directions and $a = 10\lambda$. The desired direction is chosen to be $\phi_\mathrm{s} = 0$ and the desired beam width is set to $\psi_\mathrm{s} = 5$ degree. All ϕ_m inside the desired beam width are assigned to the desired directions. However, in order to focus all energy exactly in the direction of ϕ_s, all weights in $\mathbf{C}_\mathrm{H}$ are set to 0 except the one corresponding to ϕ_s which is set to 1. This leads to the maximal gain in the desired direction, while the leakage suppression is prevented in ψ_s, leading to a broader beam. The sector L is centered around $\phi_L = 180$ degree, with $\psi_\mathrm{l} = 0$ or $\psi_\mathrm{l} = 22.5$ degree (in the former case no sector of special emphasis is considered). The weights in $\mathbf{C}_\mathrm{F}$ corresponding to the ϕ_m in the sector L are set to $c = 10$, the weights of all other undesired directions are set to 1. As can be seen, the patterns are non-regular and spiky. This comes from the random node placement and the large separation between the nodes (further discussed in Sec. 5.4). Already for a small number of cooperating nodes substantial gains can be achieved in the desired direction. While for $N = 5$ and $\psi_\mathrm{l} = 0$ the sidelobe suppression is not satisfactory, it strongly increases for increasing number of cooperating nodes. If we, in contrast to that, consider a sector of special emphasis, it can be seen that the leakage is strongly suppressed in this sector already for $N = 5$. For increasing number of cooperating nodes, the gain in the desired directions can be increased and the leakage in all undesired directions and especially in the sector can be further decreased.

5.3.1 The Signal Power - Leakage Power Trade-Off

As shown in Sec. 4.3, the weight vector $\mathbf{w}$ can be separated into a signal term $\mathbf{w}_\mathrm{s}$, shaping the signal in the desired direction, and a compensation term $\mathbf{w}_\mathrm{c}$ in an orthogonal subspace, minimizing the leakage without affecting the signal term:

$$\mathbf{w} = \mathbf{w}_\mathrm{s} + \mathbf{w}_\mathrm{c}. \tag{5.12}$$

That is, the gain in the desired direction depends on the amount of the available signal power invested into the compensation term.

By maximizing the signal-to-leakage-power ratio in Eq. (5.11), the leakage power is minimized for the resulting signal in the desired direction. Considering the transmit power - leakage power plane as introduced in Fig. 4.2 in Chap. 4, the resulting operating point would lie on the green curve (altruistic case). That is, reducing the leakage power is traded off for lower signal power in the desired direction, resulting in a smaller gain in the transmission range. Analogously to Chap. 4, this trade-off can be adjusted by accepting a higher leakage power level in order to increase the transmission range. To this end, instead of maximizing the signal-to-leakage-power ratio, we maximize the signal-to-weighted-sum-power ratio, with the weighted sum power

$$\begin{aligned} P_\vartheta &= \vartheta \cdot P_\mathrm{L} + (1-\vartheta) \cdot P_\mathrm{Tx} \\ &= \vartheta \cdot \mathbf{w}^\mathsf{H}\mathbf{F}^\mathsf{H}\mathbf{F}\mathbf{w} + (1-\vartheta) \cdot \mathbf{w}^\mathsf{H}\mathbf{w} \\ &= \mathbf{w}^\mathsf{H}\left(\vartheta \cdot \mathbf{F}^\mathsf{H}\mathbf{F} + (1-\vartheta) \cdot \mathbf{I}\right)\mathbf{w} \\ &= \mathbf{w}^\mathsf{H}\tilde{\mathbf{F}}^\mathsf{H}\tilde{\mathbf{F}}\mathbf{w}. \end{aligned} \tag{5.13}$$

The optimization problem thus becomes

$$\hat{\mathbf{w}} = \arg\max_{\mathbf{w}} \frac{\mathbf{w}^\mathsf{H}\mathbf{H}^\mathsf{H}\mathbf{H}\mathbf{w}}{\mathbf{w}^\mathsf{H}\tilde{\mathbf{F}}^\mathsf{H}\tilde{\mathbf{F}}\mathbf{w}}. \tag{5.14}$$

In contrast to the rate maximization problem in Chap. 4, only a single stream is transmitted. That is, no waterfilling among the streams is required and thus the

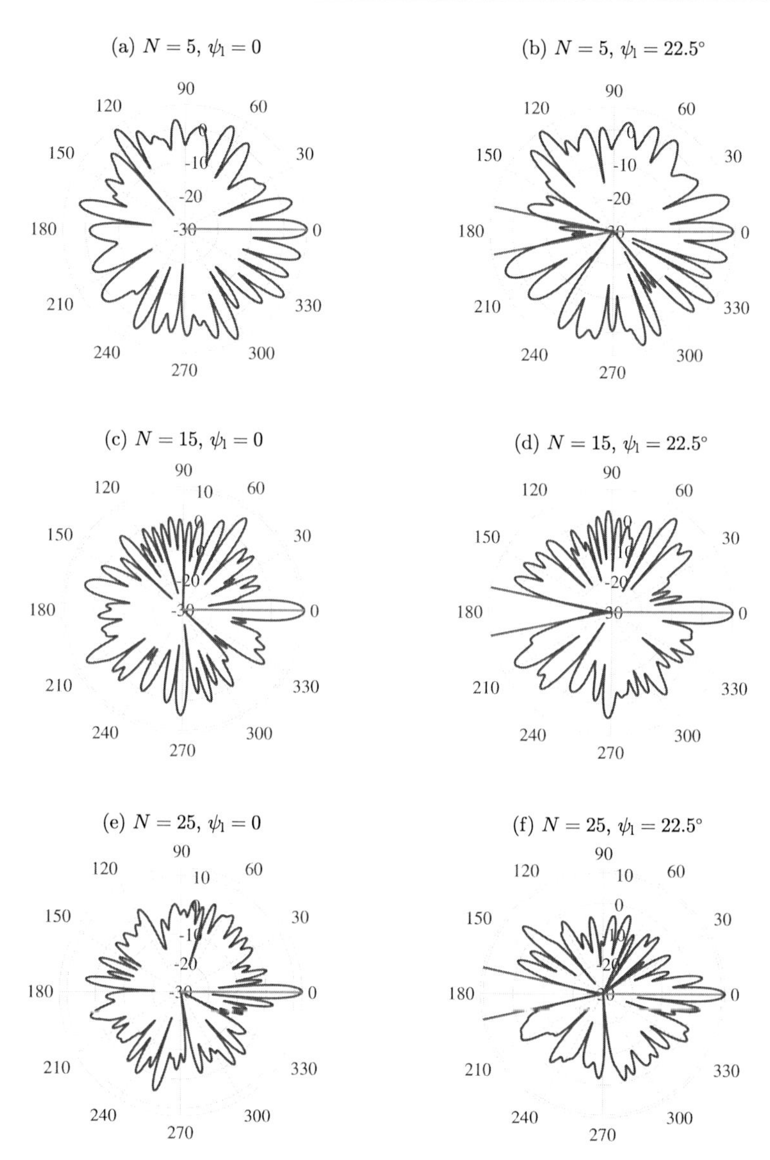

Figure 5.4.: Exemplary radiation patterns ([dBi]) of leakage based beam shaping for $N \in \{5, 15, 25\}$, $M = 3600$, $a = 10\lambda$, $\phi_s = 0$ (in green) with $\psi_s = 5°$, $c = 10$ and $\psi_l \in \{0, 22.5°\}$ centered around $\phi_L = 180°$(in red).

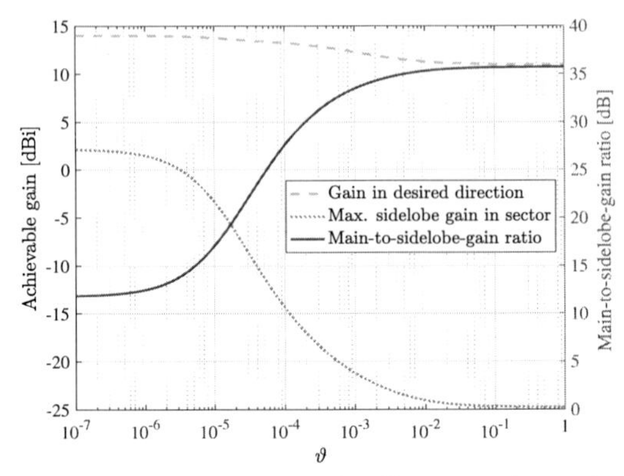

Figure 5.5.: Transmit power - leakage power trade-off corresponding to the radiation pattern in Fig. 5.4 (f) ($N = 25$, $a = 10\lambda$, $\psi_1 = 22.5°$).

optimization problem is independent of the transmit power. Only ϑ matters. Hence, the optimal $\hat{\mathbf{w}}$ can be found analogously to Eq. (5.11). The resulting $\hat{\mathbf{w}}$ can then simply be scaled to the desired transmit power.

With $\vartheta \in [0, 1]$ the trade-off between signal power and leakage power can be controlled. For $\vartheta = 0$, the leakage power is irrelevant, i.e. no leakage suppression is considered (egoistic case in Chap. 4). Hence, the signal power is simply maximized, leading to the *range maximizing solution.* For $\vartheta = 1$, the transmit power is irrelevant, leading to the optimization problem in Eq. (5.11) (altruistic case in Chap. 4). That is, the *leakage minimizing solution* is achieved. For $\vartheta \in (0, 1)$, any signal power level between coherent combining and minimal leakage power for a given transmit power can be achieved. Hence, depending on the necessary gain for the range extension, ϑ can be adjusted, resulting in the corresponding leakage power.

In Fig. 5.5 the transmit power - leakage power trade-off corresponding to the radiation pattern in Fig. 5.4(f) is illustrated by considering the achievable power gain in the desired direction, the maximal sidelobe gain in the sector L (i.e. the maximal value of the radiation pattern in sector L) and the ratio of these two gains. While with $\vartheta = 0$ the maximum possible gain can be achieved in the desired direction ($\approx$ 14 dBi), the maximal leakage level in the sector is also amplified compared to an isotropic antenna, increasing the probability of being detected and localized by hostile units. Increasing ϑ decreases the gain in the desired direction. At the same time however, the maximum sidelobe level is strongly reduced, leading to an increased main-to-sidelobe-gain ratio.

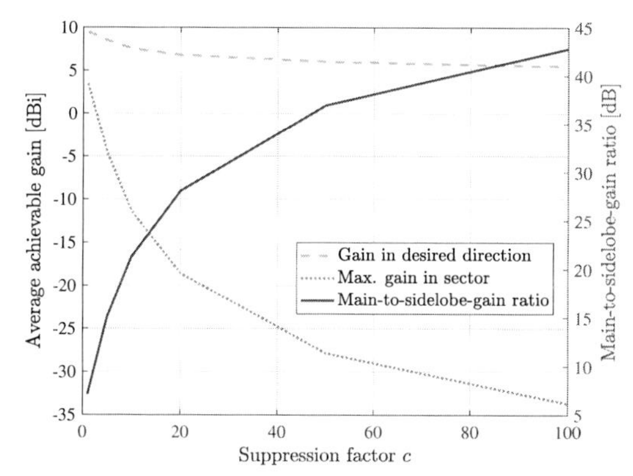

Figure 5.6.: Impact of the suppression factor c for $N = 10$, $a = 10\lambda$ and $\psi_\mathrm{l} = 22.5°$.

As for a large number of look directions M the leakage power term becomes very large compared to the transmit power term in Eq. (5.13), only a very small ϑ already shifts the focus strongly from the transmit power towards the leakage power. That is, the interesting transitions from the egoistic case to the altruistic case occur at very low ϑ. Note: Although the main-to-sidelobe-gain ratio is maximized for $\vartheta = 1$ in this example, this is not generally the case, as in the optimization the total weighted leakage power is considered and not the maximal sidelobe level in sector L.

Another way to control the trade-off between the signal power and the leakage power is the suppression factor c in the sector L. This factor strongly affects the performance of leakage based beam shaping. For a higher c, the leakage in the sector is suppressed more strongly. However, as for larger c the compensation term requires a higher share of the total transmit power, the achievable gain in the desired direction is decreased. Nevertheless, the decrease in the desired signal power is much smaller than in the leakage power. Thus, the main-to-sidelobe-gain ratio can be strongly increased. This is shown in Fig. 5.6 for $N = 10$, $a = 10\lambda$ and $\psi_\mathrm{l} = 22.5$ degree. Of course, more sophisticated weighting matrices could be applied. However, to optimize $\mathbf{C}_\mathrm{H}$ and $\mathbf{C}_\mathrm{F}$ iterative optimization would be required. As this contradicts the idea of a low complexity scheme, this is not further considered in this thesis.

Hence, the trade-off between desired signal power and leakage suppression can be controlled in two different ways: by adapting c and by varying ϑ. Choosing c large and then varying ϑ such that the desired trade-off is achieved allows for large ranges of the signal and leakage power. However, if no optimization of the trade-off is desired

N	$a = 10\lambda$	$a = 50\lambda$	$a = 100\lambda$	$a = 200\lambda$	$a = 300\lambda$
5	38	187	368	740	1106
40	42	208	413	818	1216

Table 5.1.: Average number of peaks in the radiation pattern (rounded).

or required, c can be chosen such that on average the desired gains are achieved.

5.4 Performance Evaluation

In the following, we are going to evaluate the proposed leakage based beam shaping in a setup as sketched in Fig. 5.3 for varying cluster size a, number of cooperating nodes N, and sector width ψ_l. The desired direction is set to $\phi_s = 25$ degree with $\psi_s = 5$ degree, the sector L is centered around $\phi_L = 150$ degree, and $M = 3600$ look directions are considered. As for the exemplary radiation patterns in Fig. 5.4, the weight for ϕ_s is set to 1 and all other weights for the look directions in ψ_s are set to 0. The weights for the undesired look directions are set to $c = 10$ inside the sector L and to 1 outside. For the optimization, we always consider the leakage minimizing solution, i.e. $\vartheta = 1$. 1000 Monte Carlo simulations were performed and the results averaged.

As seen in Fig. 5.4, the resulting radiation patterns are irregular and very spiky due to the randomness of the node location and their spatial separation. The number of peaks in the pattern even further increases with increasing spatial separation of the nodes. This can be seen in Tab. 5.1 which lists the average number of peaks in the radiation pattern for various side lengths a and number of nodes N. While the number of nodes N only has a small impact, the average number of peaks is strongly affected by the side length a. The increasing number of peaks in the radiation pattern strongly affect the resulting gains as we will see in the following.

For the discussion of the achievable gains of leakage based beam shaping, we firstly consider the average maximal sidelobe gain in the sector L. It is shown in Fig. 5.7(a) for a sector width of $\psi_l = 22.5$ degree and varying side length a and number of cooperating nodes N. Generally, it can be seen that the higher the number of cooperating nodes, the better is the leakage suppression as more degrees of freedom are available. However, the maximal sidelobe gain strongly increases for increasing side length a and is even above the leakage level of an isotropic radiator at large cluster size. This comes from the fact that an increased side length a results in more peaks in the radiation pattern

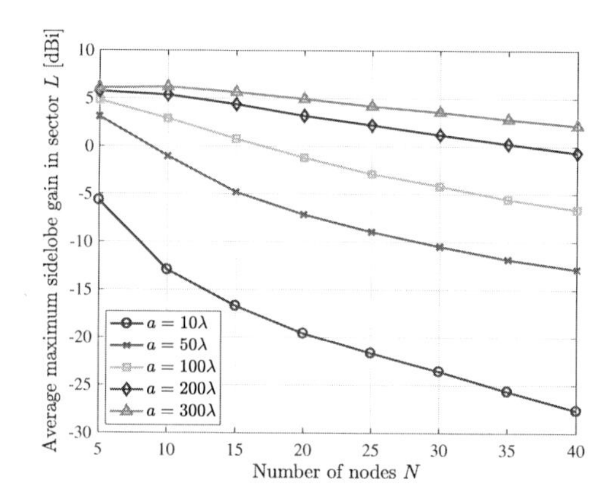

(a) Average maximal sidelobe gain in the sector L.

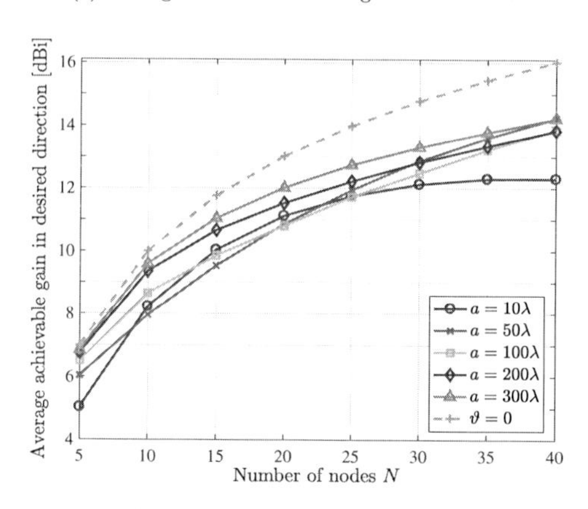

(b) Average gain in desired direction.

Figure 5.7.: Achievable gains of leakage based beam shaping in the desired direction and in the sector L for $\psi_l = 22.5°$ compared to an isotropic radiator.

(see Tab 5.1). Hence, more peaks have to be strongly suppressed in the sector L and the probability that one of these peaks is very strong (i.e. many of the individual transmit signals add up constructively in this direction) and can't be suppressed efficiently is increased. Nevertheless, for small spatial dimensions of the transmit cluster ($a = 10\lambda$), the leakage suppression works very well, already for a small amount of cooperating nodes.

In Fig. 5.7(b) the corresponding average achievable gains in the desired direction are shown. These gains depend on the fraction of the transmit power used for the compensation term. This again depends on how many peaks have to be suppressed and how well they can be suppressed, i.e. how many degrees of freedom are available. Apparently, no fixed rule can be stated considering the impact of the side length a on the gain in the desired direction. For every N in the figure the order of the achievable gains changes. The only generally valid statement can be made about the increasing gain with increasing N. At the same time, the gap between the actually achieved gain and the maximal possible gain of the range maximizing solution ($\vartheta = 0$) increases. A larger fraction of the transmit power is used for the compensation term for larger N, as the leakage can be suppressed better due to the increased degrees of freedom.

Considering the achievable main-to-sidelobe-gain ratio everything is in the expected order again as can be seen in Fig. 5.8(a). The average achievable ratio is strongly increasing with the number of nodes (higher gain in the desired direction, better leakage suppression). Already for low number of cooperating nodes very promising ratios can be achieved for low side length a. For increasing side length a, the achievable ratios are strongly decreasing due to the poor leakage suppression.

While the cluster size a has a very strong impact on the achievable performance, the impact of the sector width ψ_l is smaller. This can be seen in Fig. 5.8(b) which shows the average main-to-sidelobe-gain ratio for varying sector width ψ_l and fixed side length $a = 10\lambda$. Obviously, the gain is decreasing for large sector widths, as more leakage needs to be strongly suppressed in the sector. However, especially for a large number of cooperating nodes N, the degradation from a sector width of 11.25 degree to 90 degree is rather small. Hence, even if the direction of the hostile units is only known with a large uncertainty and thus the sector width has to be chosen big (or if multiple hostile units in a large sector have to be suppressed) still large gains in the transmission range can be achieved.

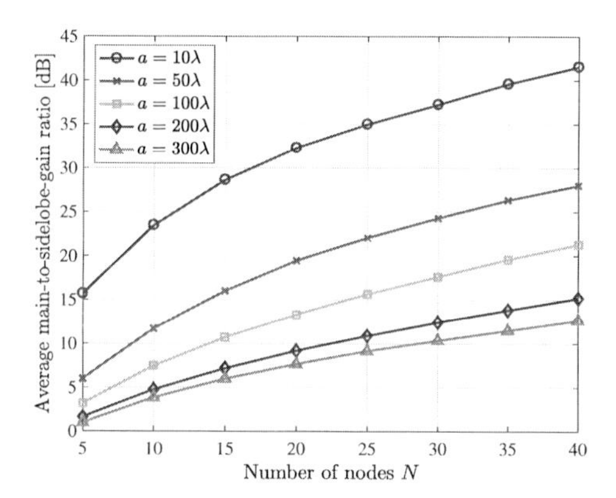

(a) Varying side length a and fixed $\psi_l = 22.5$ degree.

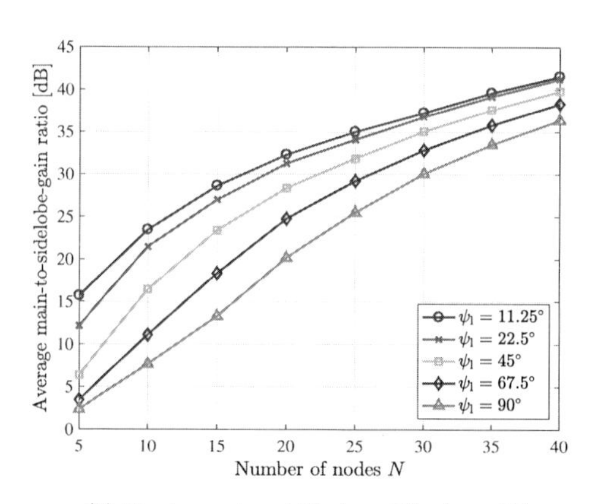

(b) Varying sector width ψ_l and fixed $a = 10\lambda$.

Figure 5.8.: Average main-to-sidelobe-gain ratio of leakage based beam shaping.

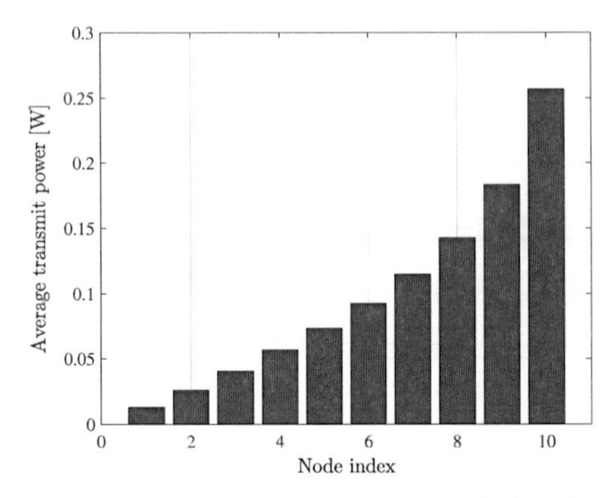

Figure 5.9.: Average sorted transmit power per node of leakage based beam shaping for $N = 10$, $a = 10\lambda$ and $\psi_l = 22.5$ degree for $P_{\mathrm{tot}} = P_{\mathrm{Tx}} = 1$ W.

5.5 Practical Considerations and Discussion

As shown above, leakage based beam shaping has a huge potential to increase the transmission range while decreasing the leakage into the network at low computational complexity. In practice however, various factors impact the performance and have to be considered. In the following, we are going to discuss several such aspects.

Transmit power Except for the range maximizing solution at $\vartheta = 0$, where only the phase at the transmitters is adapted, also the amplitudes in the weight vector are adapted in leakage based beam shaping in order to suppress the leakage. This can lead to a strongly varying transmit power per node. While the total sum transmit power of the virtual antenna array is set to the transmit power of a single node $P_{\mathrm{tot}} = P_{\mathrm{Tx}}$, that is no problem. However, if the sum transmit power of the virtual antenna array is set to $P_{\mathrm{tot}} = N \cdot P_{\mathrm{Tx}}$, it could happen that one node strongly exceeds the normal transmit power per node P_{Tx}. This can be observed in Fig. 5.9 which shows the average sorted transmit power per node of leakage based beam shaping for $N = 10$, $a = 10\lambda$ and $\psi_l = 22.5$ degree for $P_{\mathrm{tot}} = P_{\mathrm{Tx}} = 1$ W. As can be seen, the minimal and the maximal average transmit power differ by a factor bigger than 10. If the per node power constraint can not be exceeded, the weight vector has to be scaled accordingly, limiting the achievable gain in the desired direction.

N	$a = 10\lambda$	$a = 50\lambda$	$a = 100\lambda$	$a = 200\lambda$	$a = 300\lambda$
10	6.11°	1.29°	0.64°	0.32°	0.23°

Table 5.2.: Average 3dB beam width for $N = 10$ and $\psi_\text{l} = 22.5$ degree.

Beam width As seen in Sec. 5.4 the number of peaks in the radiation pattern is strongly increasing with increasing cluster size. This also impacts the 3 dB beam width of the radiation pattern as shown in Tab. 5.2 for $N = 10$ and $\psi_\text{l} = 22.5$ degree. The more peaks, the smaller is the average width of each peak. Already for a side length of $a = 50\lambda$ the average 3dB beam width decreases from over 6 degree for $a = 10\lambda$ down to 1.29 degree. That is, although large gains can be achieved in the desired direction, the location information of the destination, respectively the desired direction has to be known very accurately in order to achieve these gains.

A straight forward way to widen the beam width is to emphasize more or all desired directions in $\mathbf{C}_\text{H}$. Depending on the weighting, the beam width can be strongly increased at the price of decreased transmission range and/or higher leakage power. However, beam widening is not further evaluated in this thesis.

Location information For beam shaping, no channel state information is required. The only information needed is the relative position of the transmitting elements to each other and the position of the destination (or at least the direction to it). The relative position information can be acquired in various forms. Three eligible technologies are described in the following:

- **Satellite based localization:** To localize the nodes within the cluster, satellite based localization systems such as the global positioning system (GPS) or differential GPS can be used. In combination with measurements from inertial measurement units a location accuracy in the order of centimeters can be achieved [76].
- **Self-calibration:** In this case, the nodes try to calibrate themselves on their own using e.g. ultra-wideband (UWB) technology. In [77], an algorithm for self-calibration is suggested, with which an accuracy in the order of centimeters can be achieved.
- **External calibration:** In this case, there are various nodes far from the transmit cluster which carry out the localization process, e.g., based on angle of arrival estimation. High accuracy can be achieved with sophisticated algorithms [78].

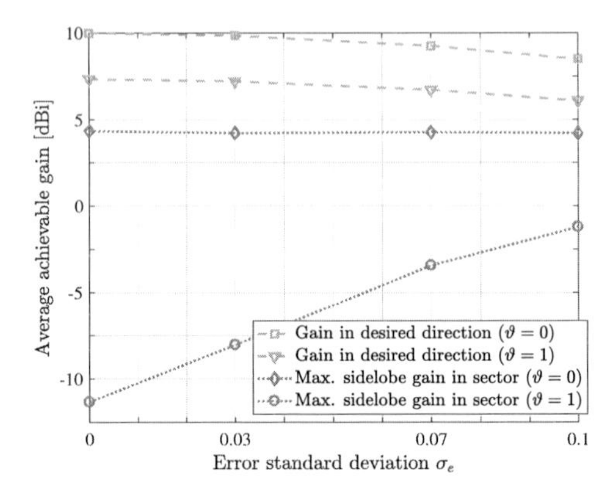

Figure 5.10.: Impact of erroneous location information on the achievable gains for $N = 10$, $a = 10\lambda$, $\psi_1 = 22.5$ degree and $c = 10$.

All these technologies can lead to high accuracy. Nevertheless, non of them can determine the location information perfectly. Thus, we are going to investigate the impact of erroneous location information in the following. To this end, we consider

$$\tilde{x}_i = x_i + e_{x,i}, \tag{5.15}$$

$$\tilde{y}_i = y_i + e_{y,i} \tag{5.16}$$

with

$$e_{x,i}, e_{y,i} \overset{\text{i.i.d}}{\sim} \mathcal{N}(0, \sigma_e^2) \quad \text{for } i = 1, \ldots, N. \tag{5.17}$$

The standard deviation of the error is set to $\sigma_e \in \{0, 0.03, 0.07, 0.1\} \cdot \lambda$. For a transmit frequency of 100 MHz (reasonable for military networks), i.e. $\lambda = 3$ m, $\sigma_e = 0.1\lambda$ corresponds to $\approx 95\%$ of the values inside ± 0.6 m of the original x and y position. For $\sigma_e = 0.03\lambda$ roughly 95% of the values are inside ± 0.2 m.

The resulting average gain in the desired direction, as well as the average maximal sidelobe gain in sector L are shown in Fig. 5.10 for the range maximizing solution ($\vartheta = 0$) and the leakage minimizing solution ($\vartheta = 1$). While the loss in the achievable gain in the desired direction is less than 2 dB for $\sigma_e = 0.1\lambda$ for both solutions, the increase of the maximal sidelobe level in the sector L for the leakage minimizing solution is drastic, even for a very low standard deviation of $\sigma_e = 0.03\lambda$. That is, the leakage suppression is very sensitive to erroneous position information. Nevertheless, it is still possible to suppress the leakage below the level of an isotropic radiator (0 dBi). Furthermore, compared to the range maximizing solution where no leakage is suppressed, still significant gains can be achieved.

Synchronization The synchronization of the nodes in time, frequency and phase is crucial. While the time and frequency synchronization are assumed to be available with a sufficient accuracy in a military MANET, accurate phase synchronization is challenging. Without proper synchronization, the performance suffers strongly, as shown in the following.

An erroneous phase synchronization has the same impact on the performance as erroneous location information. Both lead to a phase offset in the steering vector. This can be seen by considering the steering vector entry of an element i, $h_i = \exp(-j\theta_i)$, with the phase $\theta_i = \frac{2\pi}{\lambda}d_i$. An erroneous phase synchronization would lead to an offset $\Delta\theta_i$ in the steering vector, $\theta_i = \frac{2\pi}{\lambda}d_i + \Delta\theta_i$. Compared to an error in the location information in x-direction and considering $\phi = 0$, i.e. $\theta_i = \frac{2\pi}{\lambda}(x_i + \Delta x_i)$, we can see that a zero mean Gaussian distributed phase error with standard deviation $\sigma_\theta = \frac{2\pi}{\lambda} \cdot \sigma_e$ has the same impact on the performance as a zero mean Gaussian distributed location error in x-direction with standard deviation σ_e.

That is, for $\sigma_\theta = 2\pi \cdot 0.1 = 0.628$ rad (i.e. 95% of the phase errors are within ± 1.2 rad) we get the same performance degradation as for a position error with $\sigma_e = 0.1\lambda$ (see Fig. 5.10).

Scatterers Throughout the chapter we considered the radiation pattern of the antenna array for the performance evaluation. It indicates the gain in radiated power compared to an isotropic radiator in free space at large distance. Thereby, scatterers impact the predicted gains due to constructive or destructive addition of the signal at the destination. However, if the maximal gain of the radiation pattern is achieved in the desired direction, the impact of a scatterer is decreased, as its contribution is scaled according to the array pattern. That is, while the gain compared to an isotropic radiator is decreased for a constructive scatterer, it is increased for a destructive scatterer.

A scatterer could also have a significant impact on the signal power at a hostile unit. While the leakage is minimized in the sector L, a scatterer outside of the sector could reflect signal energy to the hostile units. If the reflected signal is strong enough, detection is possible. However, if the direction of the scatterers is known, multiple sectors can be realized to suppress the leakage power also in their directions.

Coupling In the system setup we explicitly assumed that all transmitting antennas are uncoupled (i.e., that they have a minimum pairwise separation of $\lambda/2$) and that they all have the same omnidirectional radiation pattern. In such a case, the total radiation pattern is simply the array pattern multiplied by the element pattern [71]. However, if antennas are spatially close (separation smaller than $\lambda/2$), they get coupled. Antenna coupling is a near field phenomenon in which a current flowing in a given antenna would induce a significant voltage on a closely located antenna [71]. If this is not considered in the optimization, antenna coupling can strongly affect the radiation pattern and decrease the performance. However, it can also lead to improved performance if utilized optimally. In [79], e.g., large gains in the spatial multiplexing are achieved if strongly coupled antennas are passively loaded (i.e. using no transmit power, just optimizing the reflection coefficient). We expect similar achievable positive effects for beam shaping.

Considering a platoon of soldiers forming a virtual antenna array at a transmit frequencies at the lower end of the VHF band ($\lambda = 10$ m), strong coupling is to be expected if they are spatially close. For high frequencies, coupling is rather unrealistic if not provoked on purpose (due to the very small necessary separation).

Coordination, clustering and data exchange The clustering of the transmitting nodes and their data exchange is rather simple. From the routing table, a node which wants to initiate a cooperative transmission knows all nodes which it can reach in one hop. Hence, it can address the required number of nodes and broadcast the message and all the necessary information to them (such as participating nodes, time table, transmit frequency, possibly a synchronization signal, ...). If necessary, they execute an array calibration, determine the weight vector and are then able to transmit with the desired pattern.

Of course a cooperation cluster could also be formed over two hops. However, this leads to a larger delays and very large clusters with all the drawbacks discussed above.

Orientation of antennas Throughout the chapter we considered the antenna pattern of all transmitting nodes to be the same. However, if, e.g., a dipole antenna of a mobile unit is not perfectly vertically positioned, its radiation pattern changes. This would have an impact on the amplitude of the received signal and not on the phase. While the leakage suppression could strongly suffer from that (if the differences in the radiation patterns of the antennas are large), the coherent addition in the desired direction is still obtained.

5.6 Conclusions

In this chapter, we introduced leakage based beam shaping, a low complexity pattern synthesis approach based on maximizing the signal-to-leakage-power ratio [18], and investigated its performance for user cooperation in military MANETs including practical implementation issues. Due to the randomly placed nodes and their possibly large separation, the patterns become irregular and spiky with sidelobes that are hard to control. Nevertheless, only one iteration in the pattern design is necessary and a closed form solution can be found in contrast to adaptive array algorithms which iteratively optimize the pattern. The sidelobes can be efficiently suppressed while achieving substantial gains in the desired direction. Hence, it leads to large gains in transmission range while reducing the probability of being detected by hostile units.

6

Quantize-and-Forward Virtual MIMO Receive Cooperation

In this chapter we consider a wireless virtual multiple-input multiple-output (MIMO) receive cooperation scheme exploiting the spatial multiplexing gain. Multiple non-cooperating sources individually transmit over multiple quantize-and-forward relays to a final destination which jointly decodes the source data streams based on the quantized observations at the relays. Such a setup is of high practical relevance in military as well as civil mobile ad hoc networks (MANETs) and wireless sensor networks in order to increase the spectral efficiency. It could reflect a direct communication between two spatially separated clusters of nodes (e.g., a mobile military unit with multiple sources transmitting to another mobile unit or the headquaters over large distances), or equivalently the communication between many independent nodes to one destination supported by multiple relays (e.g., multiple soldiers distributed in space simultaneously transmitting to their commander without a direct link). Furthermore, this setup is closely related to the downlink of user cooperation enabled traffic offloading as discussed later in Chap. 8, where multiple mobile nodes form a cluster and share a quantized version of their received signal with each other in order to jointly decode the message of independent transmitters.

In order to avoid collisions at the final destination only one relay forwards its quantized observation at a time. The achievable rate on the link to the final destination (further called backhaul rate) as well as the channel access time thereby determine the quantization rate of the relays, which eventually impacts the final decoding rate. Furthermore, the total duration of the second hop strongly affects the system throughput. Therefore, the backhaul resources need to be assigned carefully in order to optimize the system performance, leading to an involved non-convex optimization problem.

The goal of this chapter is to provide simple and robust resource allocation algorithms, which are well suited for military applications as well as for the traffic offloading in urban hotspots discussed in Chap. 8. Hence, as military equipment tends to boost long life time cycles, the proposed algorithms should be implementable as an incremental update on existing hardware (half-duplex nodes with single-antennas), and should require only low computational complexity.

In this context, we devise simple relay selection schemes and compare them to multi-start gradient search and to equal resource allocation for all relays. The relay selection is either based on their receive signal-to-noise ratio (SNR) or on their signal-to-quantization-noise ratio (SQNR). We show that for a predetermined second hop duration, large gains can be achieved by choosing an optimized subset of the relays. Further optimizing the channel access times only leads to minor additional gains. However, to achieve these gains, it is crucial to know the optimal number of relays in the subset and the total resources to allocate in the backhaul access phase. If these numbers are not known in advance, it is computationally demanding to find them. Hence, these schemes are only suited for quasi stationary setups, where the number of participating nodes and the channel characteristics change slowly. To this end, based on the observations made with the relay selection schemes, we propose the *cascade resource allocation* algorithm which inherently determines the optimized subset of relays and the channel access times. It leads to promising results compared to gradient search at low computational complexity.

This chapter is mainly based on our work published in [35] and [38]. The remainder is structured as follows. In Sec. 6.1 related work is reviewed. In Sec. 6.2 the system model is presented and the optimization problem derived, and in Sec. 6.3 the resulting resource allocation problem is discussed. In Sec. 6.4 the relay selection schemes are introduced and evaluated and in Sec. 6.5 the cascade algorithm is introduced and evaluated. Sec. 6.6 finally concludes the chapter.

6.1 Related Work

Related work has primarily been performed in the context of distributed MIMO receivers and cloud radio access networks (CRANs) (e.g., [80–84]). In [80] and [81], a two user setup with orthogonal backhaul links is considered. While [80] analyzes the performance of quantize-and-forward and compress-and-forward relays for given backhaul link rates, [81] considers the same problem with random fluctuations on the backhaul. Different to quantize-and-forward, compress-and-forward includes distributed source

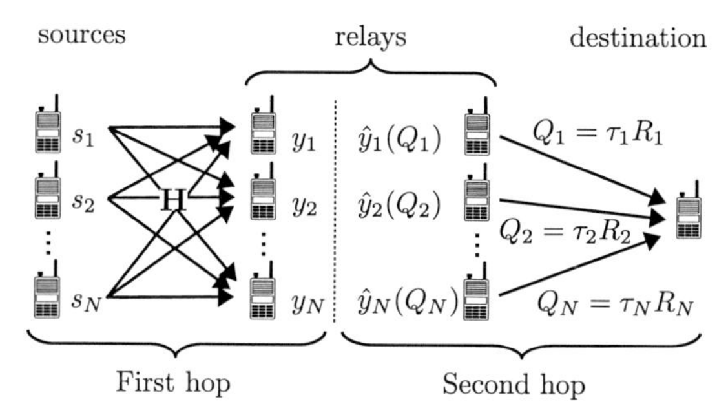

Figure 6.1.: System setup.

coding, leading to much higher complexity in the encoding and decoding and requiring full channel state information at the relays. In [82], a CRAN uplink scenario with orthogonal backhaul links is considered and the optimal quantization noise levels for different compress-and-forward schemes are derived. Similarly, but for a shared backhaul with a sum-capacity constraint, in [83] the optimal quantization noise level for each relay is derived at high signal-to-quantization-noise ratio for quantize-and-forward and compress-and-forward. Low complexity algorithms are presented for the quantization noise level design. In [84], the setups of [82] and [83] are extended to multi-antenna terminals and investigated in terms of their achievable rate regions.

Different to the setup at hand, in [80–84] the second hop is either already considered orthogonalized (i.e. no resource allocation is necessary [80–82, 84]), or with a sum capacity constraint [83, 84]. However, especially in wireless virtual MIMO setups non-orthogonal backhaul links as well as individual rate constraints are of high practical relevance, as each link interferes with the others and experiences independent shadowing and fading.

6.2 System Model and Problem Formulation

Throughout the chapter, we consider the system model as shown in Fig. 6.1, with N sources, N relays and one final destination. Although with a higher number of relays, the performance for the N sources could be increased due to the additional available diversity, the number of sources is considered such that theoretically the maximal possible spatial multiplexing gain can be achieved with the available relays.

Motivated by the typical downward compatibility and evolutionary growth requirements of military networks, we consider all nodes to be half-duplex with a single antenna. The half duplex assumption imposes quite stringent constraints on the feasible traffic patterns. Specifically, we assume that:

- Relays cannot receive on the first hop while transmitting on the second hop.
- If any one relay transmits, non of the other relays can receive simultaneously.

We furthermore assume that the relays transmit one at a time (thus the destination is not required to resolve collisions). The allocated time to relay i is denoted by τ_i. For the clarity of exposition we normalize without loss of generality all time intervals to the duration of 1 channel use of the first hop.

The wireless channel between the sources and the relays is denoted by $\mathbf{H} \in \mathbb{C}^{N \times N}$, with $h_{i,j}$ the channel from source j to relay i. The transmit symbol of source i is assumed to be $s_i \sim \mathcal{CN}(0, P_{\text{Tx}})\ \forall\, i \in \{1, \ldots, N\}$. The received signal of the relays is then a superposition of all transmit signals

$$y_i = \sum_{j=1}^{N} h_{i,j} \cdot s_j + z_i, \tag{6.1}$$

where z_i denotes the additive white Gaussian noise $w_i \sim \mathcal{CN}(0, \sigma_w^2)$ plus potential circularly complex Gaussian distributed interference. The variance of the received signal is thus given as

$$\sigma_{y_i}^2 = P_{\text{Tx}} \sum_{j=1}^{N} |h_{i,j}|^2 + \sigma_{z_i}^2. \tag{6.2}$$

The backhaul link from relay i to the destination is assumed to support rate R_i bit per first hop channel use[1] (bpfhcu) with negligible probability of error (i.e. relay i can transmit R_i information bit to the destination in the time which one channel use takes on the first hop). If we let τ_{tot} be the normalized duration of the second hop, the *effective quantization rate* of the relays is given as $Q_i = \tau_i \cdot R_i$, with $\sum_{i=1}^{N} \tau_i = \tau_{\text{tot}}$ and $\tau_i \geq 0\ \forall\, i \in \{1, \ldots, N\}$. Considering a vector quantizer at the relays, the quantization noise is additive and zero-mean Gaussian distributed ([85], Chap. 10). The *achievable decoding rate* R^{QF} at the destination in bits per first hop channel use is thus given as

$$R^{\text{QF}} = \log_2 \det\left(\mathbf{I} + (\mathbf{D}_{\text{z}} + \mathbf{D}_{\text{q}})^{-1} \mathbf{\Lambda}_{\text{s}}\right), \tag{6.3}$$

[1] This corresponds to a normalization of the spectral efficiency with respect to the first hop bandwidth and thus allows to consider different bandwidths respectively symbol durations on the first and second hop.

with $\mathbf{I}$ the identity matrix, the signal covariance matrix $\mathbf{\Lambda}_\mathrm{s} = P_\mathrm{Tx} \cdot \mathbf{H}\mathbf{H}^\mathrm{H}$ and the two diagonal covariance matrices $\mathbf{D}_\mathrm{z} = \mathrm{diag}\left(\sigma^2_{z_1}, \ldots, \sigma^2_{z_N}\right)$ and $\mathbf{D}_\mathrm{q} = \mathrm{diag}\left(\sigma^2_{q_1}, \ldots, \sigma^2_{q_N}\right)$ of the interference plus noise and the quantization noise respectively.

$$\sigma^2_{q_i} = \frac{\sigma^2_{y_i}}{2^{Q_i} - 1} = \frac{\sigma^2_{y_i}}{2^{\tau_i R_i} - 1} \tag{6.4}$$

denotes the quantization noise variance of relay i, which depends on the effective quantization rate Q_i and thus on τ_i.

In total, a 2-hop transmission cycle comprises $1+\tau_\mathrm{tot}$ time units. Hence, the resulting *overall throughput* R_MS of the system is given as

$$R_\mathrm{MS} = \frac{1}{1+\tau_\mathrm{tot}} \cdot R^\mathrm{QF}. \tag{6.5}$$

Thereby, τ_tot has two opposing effects on the throughput R_MS. For increasing τ_tot, the achievable decoding rate at the final destination R^QF is increasing (if the τ_i are reasonably assigned), but the factor in front of R^QF is decreased, due to the longer second hop. That is, the trade-off leading to the maximal throughput has to be found. This optimization problem can be stated as

$$\begin{aligned} [\hat{\tau}_1, \ldots, \hat{\tau}_N] &= \arg\max_{\tau_1, \ldots, \tau_N} R_\mathrm{MS} \\ \text{s.t.} \quad & \tau_i \geq 0 \; \forall i \in \{1, \ldots, N\}, \end{aligned} \tag{6.6}$$

which inherently yields the optimal $\tau_\mathrm{tot} = \sum_{i=1}^N \tau_i$.

Note: Different to [83] where the quantization noise level is optimized for full-duplex relays (i.e. $\tau_\mathrm{tot} = 1$) under a sum capacity constraint C_sum (i.e. all Q_i could take on any value in $[0, C_\mathrm{sum}]$ as long as $\sum_i Q_i \leq C_\mathrm{sum}$), τ_tot is variable in our setup and the quantization rates are limited to $Q_i \in [0, \tau_\mathrm{tot} \cdot R_i]$, which is different for every relay.

6.3 Backhaul Resource Allocation

The optimization problem in Eq. (6.6) can be approached with the Karush-Kuhn-Tucker (KKT) conditions. Reformulating the inequality constraint $\tau_i \geq 0$, $\forall i \in$

$\{1,\ldots,N\}$ into $-\tau_i \leq 0$, $\forall i \in \{1,\ldots,N\}$, we get the Lagrange function as

$$\mathcal{L}(\boldsymbol{\tau},\boldsymbol{\mu}) = R_{\text{MS}} + \sum_{i=1}^{N} \mu_i \tau_i, \tag{6.7}$$

and the KKT conditions as

$$\begin{aligned} \nabla\mathcal{L}(\boldsymbol{\tau},\boldsymbol{\mu}) &= 0 && (6.8)\\ -\tau_i &\leq 0\,\forall i \in \{1,\ldots,N\} && (6.9)\\ \mu_i &\geq 0\,\forall i \in \{1,\ldots,N\} && (6.10)\\ \mu_i\tau_i &= 0\,\forall i \in \{1,\ldots,N\}. && (6.11)\end{aligned}$$

The partial derivatives of the gradient are given as

$$\frac{\partial\mathcal{L}(\boldsymbol{\tau},\boldsymbol{\mu})}{\partial\tau_i} = \frac{\partial R_{\text{MS}}}{\partial\tau_i} + \mu_i. \tag{6.12}$$

Hence, a relay is either operating (i.e. $\tau_i > 0$), leading to $\mu_i = 0$ (from Eq. (6.11)) and thus $\partial R_{\text{MS}}/\partial\tau_i = 0$ (from Eq. (6.8)), or it is inoperative (i.e. $\tau_i = 0$), leading to $\mu_i \geq 0$ and thus $\partial R_{\text{MS}}/\partial\tau_i \leq 0$. That is, there is no gain in the achievable rate if incremental resources are reassigned from one relay to another and a local maximum is achieved.

The partial derivatives of R_{MS} can be found as

$$\frac{\partial R_{\text{MS}}}{\partial\tau_i} = R^{\text{QF}} \cdot \frac{\partial}{\partial\tau_i}\frac{1}{1+\sum_{j=1}^{N}\tau_j} + \frac{1}{1+\sum_{j=1}^{N}\tau_j} \cdot \frac{\partial}{\partial\tau_i}R^{\text{QF}}, \tag{6.13}$$

with

$$\frac{\partial}{\partial\tau_i}\frac{1}{1+\sum_{j=1}^{N}\tau_j} = -\frac{1}{(1+\sum_{j=1}^{N}\tau_j)^2} \tag{6.14}$$

and

$$\begin{aligned} \frac{\partial R^{\text{QF}}}{\partial\tau_i} &= \frac{\partial}{\partial\tau_i}\left(\log_2\det\left(\mathbf{D}_\text{q}+\mathbf{D}_\text{z}+\boldsymbol{\Lambda}_\text{s}\right) - \log_2\det\left(\mathbf{D}_\text{q}+\mathbf{D}_\text{z}\right)\right)\\ &= \operatorname{tr}\Big(\left(\mathbf{D}_\text{q}+\mathbf{D}_\text{z}+\boldsymbol{\Lambda}_\text{s}\right)^{-1}\frac{\partial}{\partial\tau_i}\mathbf{D}_\text{q}\Big) - \operatorname{tr}\Big(\left(\mathbf{D}_\text{q}+\mathbf{D}_\text{z}\right)^{-1}\frac{\partial}{\partial\tau_i}\mathbf{D}_\text{q}\Big)\\ &= \Big(\left(\left(\mathbf{D}_\text{q}+\mathbf{D}_\text{z}+\boldsymbol{\Lambda}_\text{s}\right)^{-1}\right)_{i,i} - \left(\left(\mathbf{D}_\text{q}+\mathbf{D}_\text{z}\right)^{-1}\right)_{i,i}\Big) \cdot \frac{R_i\sigma_{y_i}^2 2^{\tau_i R_i}}{(2^{\tau_i R_i}-1)^2}, \end{aligned} \tag{6.15}$$

where the last step follows from the fact that in $\partial/\partial\tau_i\,\mathbf{D}_\text{q}$ only the element in row i and column i (denoted by $(\cdot)_{i,i}$) is unequal to zero. Due to the coupling of the partial derivatives of R_{MS} in all the τ_i, we are not aware of an elementary solution to the problem. Furthermore, as can be seen with numerical examples, multiple local optima may exist. That is, the problem is non-convex. Hence, either a computationally expensive optimization algorithm is applied such as a multi-start gradient search (still not guaranteed to find the optimal solution), or a suboptimal but low complexity solution is considered.

6.4 Relay Selection

In the following we focus on two low complexity relay selection schemes. They are compared to the reference approach of equal time shares for all relays and to a gradient search with multiple initializations. For all of them, the channel access times τ_i are allocated for a fixed τ_{tot} and then various τ_{tot} are considered in order to illustrate the system performance in dependence of the second hop duration.

Reference approach: equal time allocation As a baseline we consider equal time allocation for all relays, i.e. $\tau_i = \tau_{\text{tot}}/N \ \forall\, i \in \{1, \dots, N\}$. Not taking the information content of the observations of the relays and the backhaul rates into account can lead to highly suboptimal results for setups with strong variations in backhaul rates and signal strength at the relays.

First hop relay selection (1RS) Depending on the backhaul rates and the available resources, it is crucial to only consider a subset of relays for the resource allocation. To keep the approach as simple as possible we just select N_{o} operating relays based on their receive $\text{SNR}_i = \frac{\lambda_{i,i}}{\sigma^2_{z_i}}$, with $\lambda_{i,i}$ the element of $\mathbf{\Lambda}_{\text{s}}$ in row i and column i, and assign equal $\tau_i = \tau_{\text{tot}}/N_{\text{o}}$ to them. The N_{o} as well as the τ_{tot} are thereby considered to be pre-determined. That is, the only necessary information for the relay selection is the SNR of the first hop. The crucial part however, is to choose τ_{tot} and N_{o} in advance (e.g., based on experience).

Note: For $N_{\text{o}} = N$, it corresponds to the reference approach of equal time-shares for all relays. For $N_{\text{o}} = 1$, no virtual MIMO is applied and simply the best relay is selected based on its SNR.

Second hop relay selection (2RS) Similar to the previous approach, the second hop relay selection considers only a subset of N_{o} operating relays. However, as the backhaul rates might have a significant impact on the performance, the relays are chosen based on their $\mathrm{SQNR}_i = \frac{\lambda_{i,i}}{\sigma_{z_i}^2 + \sigma_{q_i}^2}$, with $\tau_i = \tau_{\mathrm{tot}}/N_{\mathrm{o}}$. That is, both hops are incorporated into the choice and relays with high backhaul rate but similar received signal power as the others are favored. Again, the N_{o} as well as the τ_{tot} are considered to be predetermined. Additional to the SNRs also the backhaul rates have to be known for the choice. For $N_{\mathrm{o}} = N$ it also corresponds to the reference approach.

Gradient search The optimal solution of Eq. (6.6) can be found, e.g., by a gradient search with sufficient initializations at the price of strongly increased computational effort. Nevertheless, as a reference for the proposed resource allocation schemes, a gradient search which optimizes R^{QF} for a given τ_{tot} is considered in the evaluations as well.

Complexity While the gradient search requires in each step the evaluation of N partial derivatives which either include inverses of full matrices (analytical gradient) or determinants of full matrices (numerical gradient), both with high computational complexity, the computational complexity of the relay selection approaches is extremely low (determine N SNRs, respectively SQNRs and choose the strongest relays). Furthermore, the relay selection schemes only require the $\lambda_{i,i}$ and R_i (for the second hop relay selection), while the gradient approach requires the full knowledge of $\boldsymbol{\Lambda}_{\mathrm{s}}$ and the R_i. Hence, more data needs to be exchanged.

6.4.1 Numerical Evaluations

In the following simulations we consider a setup as described in Section 6.2 with $N = 10$ sources and quantize-and-forward relays. On the first hop, the channel coefficients from all sources to a given relay are considered i.i.d. complex Gaussian. We define $\overline{\mathrm{SNR}}_i$ as the average SNR at relay i if all sources are transmitting, and distinguish two relay deployment scenarios:

- Relay cluster: The relays are spatially close compared to the source-relay distances and thus have similar average receive SNR, $\mathbf{snr} = [\overline{\mathrm{SNR}}_1, \ldots, \overline{\mathrm{SNR}}_N]^{\mathsf{T}} = [13, 13.22, \ldots, 15]^{\mathsf{T}}$ dB in vector form.
- Scattered relays: The relays are spatially dispersed and thus have substantially different average receive SNR of $\mathbf{snr} = [\overline{\mathrm{SNR}}_1, \ldots, \overline{\mathrm{SNR}}_N]^{\mathsf{T}} = [2, 4, \ldots, 20]^{\mathsf{T}}$ dB in vector form.

	Weak shadowing	Strong shadowing
clustered relays	I	IV
scattered relays	II	III

Table 6.1.: Operation regimes.

For all simulations in this chapter we don't consider external interference and set $\sigma_w^2 = 1$, i.e. $\sigma_{z_i}^2 = \sigma_w^2 = 1 \ \forall \ i \in \{1, \ldots, N\}$. With $P_{\text{Tx}} = 1$, the resulting channel matrix is then determined as

$$\mathbf{H} = \text{diag}(\sqrt{\mathbf{snr}/N}) \cdot \mathbf{F}, \tag{6.16}$$

with $\mathbf{F}$ the fading matrix with i.i.d. elements $\sim \mathcal{CN}(0, 1)$.

On the second hop, all the rates R_i are considered to be statistically independent random variables and we distinguish two destination shadowing scenarios (remember that the rates on the second hop are given in bit per first hop channel use (bpfhcu)):

- Weak destination shadowing: The variance in the shadowing is small and the R_i are uniformly drawn from the interval $[4.5, 5.5]$ bpfhcu.
- Strong destination shadowing: The variance in the shadowing is large and the R_i are uniformly drawn from the interval $[0, 10]$ bpfhcu.

Combining the different first hop and second hop scenarios we end up with the four typical operation regimes shown in Table 6.1. For each setup 1000 Monte-Carlo simulations are conducted.

In **regime I** with similar SNRs and weakly varying backhaul rates, the results and conclusions are similar to regime II, on which we will focus in the sequel.

Regime II features strongly varying SNRs and similar backhaul rates. In Fig. 6.2 the average achievable decoding rate at the final destination R^{QF} in Eq. (6.3), is shown for the first hop relay selection, for the gradient search, as well as for the co-located MIMO case, i.e. without any quantization noise. While huge degeneration of the achievable rates can be observed for all schemes compared to the co-located MIMO case at low τ_{tot} (due to the strong quantization noise), the performance is improving for increasing τ_{tot}. However, the performance of the relay selection scheme is limited by its maximal number of spatial degrees of freedom which is determined by the number of operating relays N_{o}. This is clearly visible in Fig. 6.2. The curves for the different N_{o} only achieve a certain plateau and can not improve anymore for higher τ_{tot}. The $\sigma_{q_i}^2$ are already much smaller than the thermal noise σ_w^2 and thus do not

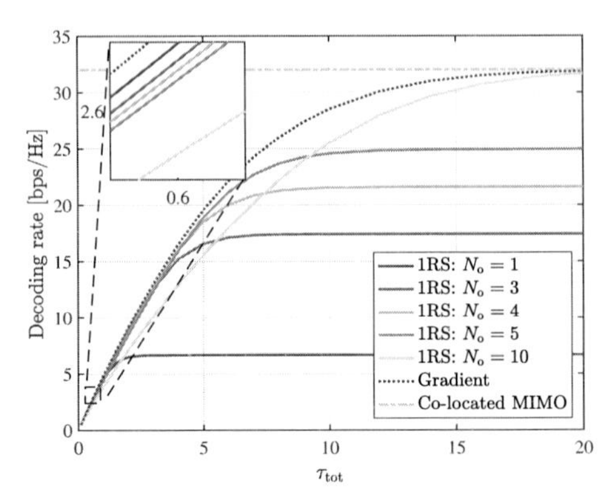

Figure 6.2.: Average decoding rate R^{QF} in regime II with $\mathbf{snr} = [2, 4, \ldots, 20]$ dB and $R_i \sim \mathcal{U}(4.5, 5.5)$ bpfhcu for the first hop relay selection scheme.

affect the performance significantly anymore. Obviously, the higher N_o, the higher is the achievable performance if τ_{tot} is large enough, and eventually the co-located MIMO bound is achieved for $N_o = 10$. However, for lower τ_{tot} (i.e. lower resulting SQNRs) a lower N_o is beneficial. Hence, it is best to only serve a subset of the relays at low SQNR. As expected, the gradient search leads to the best performance for all τ_{tot}, as it adaptively chooses the optimal number of relays and additionally optimizes the assignment of the τ_i.

The curves for the second hop relay selection are almost identical and therefore not shown. As the variance in the backhaul rates is small, the best relays are mainly determined by the first hop SNR. Hence, for similar backhaul rates, first hop relay selection is sufficient.

The corresponding total average throughput R_{MS} in Eq. (6.5) is shown in Fig. 6.3. The trade-off between high R^{QF} and long backhaul access time can be clearly observed for all curves. However, depending on the number of operating relays N_o, the peak performance is achieved at a different τ_{tot} and the width of the curves varies strongly. This comes from the fact, that for smaller N_o the plateau is achieved at lower τ_{tot}. As soon as the plateau is reached, the performance drops strongly. Eventually, the R_{MS} of all schemes will tend to 0 as τ_{tot} grows large. The best peak performance in this setup is achieved for $N_o = 4$, getting close to the performance of the gradient search. Although a large loss can be observed compared to the co-located MIMO case (due to the quantization noise and the smaller number of degrees of freedom), large gains

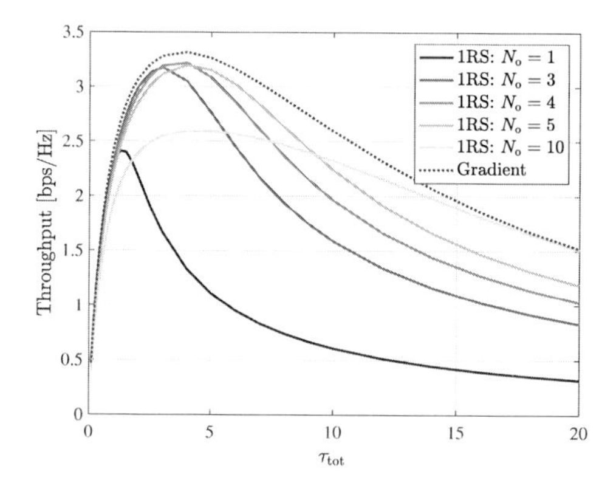

Figure 6.3.: Total average throughput R_{MS} in regime II with $\mathbf{snr} = [2, 4, \ldots, 20]$ dB and $R_i \sim \mathcal{U}(4.5, 5.5)$ bpfhcu for the first hop relay selection scheme.

can be achieved compared to $N_o = 1$ and $N_o = 10$. That is, applying a virtual MIMO scheme based on simple relay selection can strongly increase the performance in such a setup if τ_{tot} and N_o are reasonably chosen.

In **regime III** with strongly varying SNRs ($\mathbf{snr} = [2, 4, \ldots, 20]^\mathsf{T}$ dB) and large differences in the backhaul rates ($R_i \sim \mathcal{U}(0, 10)$ bpfhcu), the backhaul rates R_i have a significant impact on the performance. This can be seen in Fig. 6.4 which shows the total average throughput R_{MS} of both relay selection schemes for their specific best average N_o ($N_o = 5$ for the first hop relay selection, respectively $N_o = 3$ for the second hop relay selection), as well as for the reference schemes. Although the first hop relay selection still leads to significant gains with the optimal N_o compared to $N_o = 10$, there is a large loss compared to the second hop relay selection. That is, the backhaul rates R_i have a strong impact on the performance. A node with high SNR but low backhaul rate is wasting a lot of resources compared to a node with slightly lower SNR but much higher backhaul rate. Hence, the backhaul rates should be included in the relay selection if the variance among them is large.

Note that besides the gain in the maximal total average throughput, also large gains considering the delay can be achieved at the same performance. In contrast to the first hop relay selection, where the peak performance is achieved at $\tau_{\text{tot}} = 3$, the second hop relay selection only requires $\tau_{\text{tot}} = 0.7$ for the same throughput.

Apparently, the second hop relay selection is very efficient, getting close to the gra-

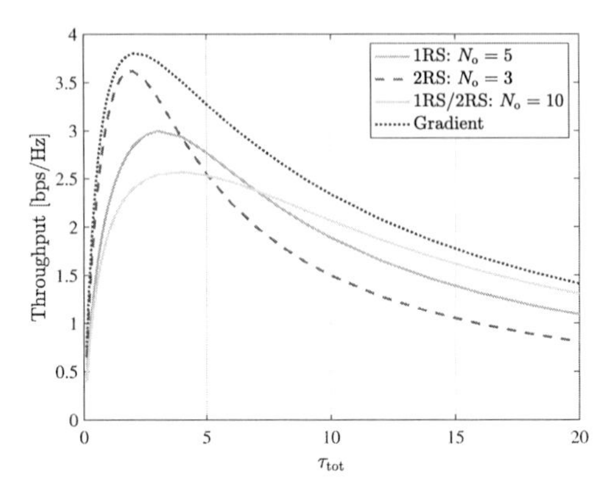

Figure 6.4.: Total average throughput R_{MS} in regime III with $\mathbf{snr} = [2, 4, \ldots, 20]$ dB and $R_i \sim \mathcal{U}(0, 10)$ bpfhcu for both relay selection schemes.

dient search performance. The most important gains are already achieved by choosing the best relays, and only minor additional gains are achieved with the optimal τ_i. This is also visualized in Fig. 6.5, where for one specific channel realization the instantaneous τ_i are shown for both relay selection schemes, the gradient search and the reference approach at their respective peak performance. The title of the figure also shows the corresponding instantaneous total throughput. While first hop relay selection achieves its best performance for $N_o = 2$, the second hop relay selection chooses the same $N_o = 4$ relays as the gradient search, achieving large gains compared to the reference approach and the first hop relay selection. The gradient search in the end only achieves minor additional gains by optimizing the τ_i, although especially for τ_3 and τ_9, the differences are quite significant.

Except for the instantaneous total throughput in Fig. 6.5, only averaged values have been considered so far. Fig. 6.6 shows the empirical CDF of the instantaneous total throughput of the two relay selection schemes if always their average best τ_{tot} and N_o are chosen (i.e. for $\tau_{tot} = 3$ and $N_o = 5$ for the first hop relay selection scheme and $\tau_{tot} = 2$ and $N_o = 3$ for the second hop relay selection scheme). It furthermore shows the empirical CDF of the two relay selection schemes and the gradient search if always the maximal instantaneous total throughput is considered (i.e. for varying τ_{tot} and N_o), as well as the corresponding empirical CDFs for the case where only one relay is chosen randomly (conventional relaying) and the reference case of equal resource allocation for all relays. It impressively visualizes how valuable it is to apply virtual MIMO, and

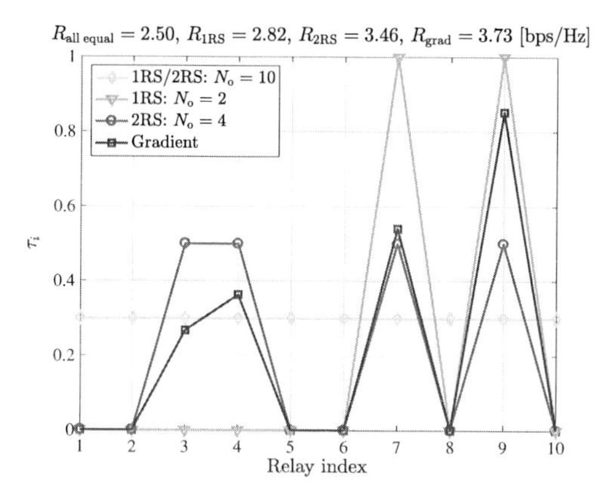

Figure 6.5.: Instantaneous τ_i at the respective peak performance for one specific channel realization in regime III with $\mathbf{snr} = [2, 4, \ldots, 20]$ dB and $R_i \sim \mathcal{U}(0, 10)$ bpfhcu.

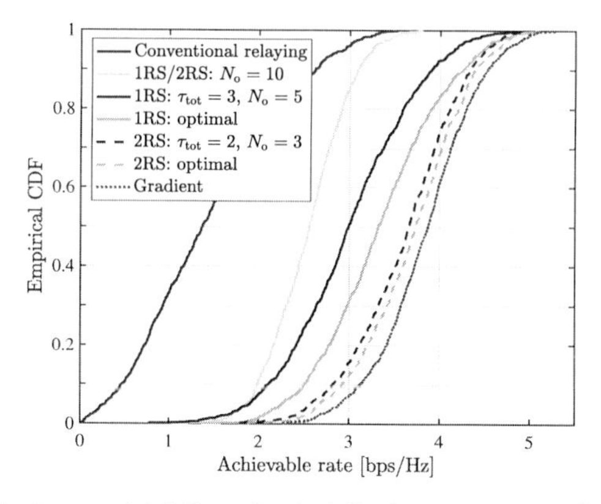

Figure 6.6.: Instantaneous total throughput at the instantaneous optimal τ_{tot} and N_o, and at the average optimal τ_{tot} and N_o in regime III with $\mathbf{snr} = [2, 4, \ldots, 20]$ dB and $R_i \sim \mathcal{U}(0, 10)$ bpfhcu.

how large the gains with a proper resource allocation can be. Furthermore, it can be seen that for the second hop relay selection scheme only minor gains can be achieved by always choosing the optimal τ_{tot} and N_{o}. Hence, considering a fixed number of operating relays and a fixed length for the second hop is reasonable if chosen wisely. This could strongly simplify the protocol in practice, as N_{o} and τ_{tot} could be set initially, e.g., based on experience, and then considered constant as long as the setup does not change significantly. For the first hop relay selection scheme, the achievable gain is much higher. That is, it is more sensitive to N_{o} and τ_{tot}. This comes from the strongly varying backhaul rates, which are not incorporated into the relay selection of the first scheme, but have a high impact on the performance.

Of course, not only the variance of the backhaul rates strongly affects the performance, but also their level. For high backhaul rates, much better quantization can be achieved with equal delay (τ_{tot}) and thus, the throughput is strongly improved. This is discussed in the context of the evaluation for the cascade algorithm in the next section.

Regime IV: As the main influence factor for the relay selection is again the backhaul rates, the characteristics of the results of regime IV are very similar to the results of regime III and are thus omitted as well.

6.5 Cascade Resource Allocation

To achieve the gains of relay selection described in the previous section, the size of the subset of the operating relays which forward their observation, as well as the total duration of the second hop have to be known in advance, or alternatively have to be found by exhaustive search. In the following, we present a low complexity resource allocation algorithm which inherently determines the subset of relays as well as the channel access times τ_i and thus τ_{tot}.

Cascade algorithm In order to identify the relay stations to consider, we propose to add one relay after the other to the subset of operating relays, as long as the throughput R_{MS} in Eq. (6.5) can be increased. The relays are thereby considered in descending order of relevance to the rate R^{QF}. We measure the relevance of relay i in terms of the partial derivative of R^{QF} with respect to the channel access time τ_i at the point

Algorithm 2 Cascade resource allocation

1: Sort relays in descending order of $\frac{\partial R^{\mathrm{QF}}}{\partial \tau_i}\Big|_{\tau_j=0\ \forall\ j\in\{1,\dots,N\}}$
2: Initialization: $\boldsymbol{\tau} = [\tau_1, \dots, \tau_N] = [0, \dots, 0]$
3: $R_{\mathrm{MS}} = 0$
4: **for** $k = 1$ **to** N **do**
5: $\quad \tau_k = \log_2(1 + \mathrm{SNR}_k)/R_k$
6: $\quad R_{\mathrm{MS,tmp}} = \frac{1}{1+\mathrm{sum}(\boldsymbol{\tau})} R^{\mathrm{QF}}(\boldsymbol{\tau})$
7: $\quad$ **if** $R_{\mathrm{MS,tmp}} > R_{\mathrm{MS}}$ **then**
8: $\qquad R_{\mathrm{MS}} = R_{\mathrm{MS,tmp}}$
9: $\quad$ **else**
10: $\qquad \tau_k = 0$
11: $\qquad$ break for loop
12: $\quad$ **end if**
13: **end for**

$\tau_1 = \tau_2 = \dots = \tau_N = 0$, i.e.

$$\frac{\partial R^{\mathrm{QF}}}{\partial \tau_i}\Bigg|_{\tau_j=0\ \forall\ j\in\{1,\dots,N\}} = \frac{\mathrm{SNR}_i}{1 + \mathrm{SNR}_i} \cdot R_i. \tag{6.17}$$

The derivation of this partial derivative is provided in App. A.5. Recall that R_i is the broadcasting rate of relay i (see Sec. 6.2). We propose to allocate to relay i the time

$$\tau_i = \log_2(1 + \mathrm{SNR}_i)/R_i, \tag{6.18}$$

such that the effective quantization rate is $Q_i = \log_2(1+\mathrm{SNR}_i)$. Here, $\mathrm{SNR}_i = \lambda_{i,i}/(\sigma^2_{z_i})$ denotes the SNR of relay station i, with $\lambda_{i,i}$ the signal power and $\sigma^2_{z_i}$ the interference plus noise power. For each potential relay station to be added, the throughput in Eq. (6.5) is evaluated. Once it decreases compared to the previous subset of operating relays or if all relays are considered already, the resource allocation is stopped. The algorithm is summarized in Alg. 2.

This procedure is motivated as follows. The relays are chosen in order of the strongest increase in R^{QF} if no relay is considered yet. Although this criterion only guarantees optimality for the choice of the first relay station (as for all succeeding relays the partial derivatives change), the succeeding relays are still a reasonable choice. From Eq. (6.17) we see that they have a strong broadcasting rate R_i and can thus provide large quantization rates Q_i while requiring only limited resources τ_i. Furthermore, permanently adapting the selection criterion would strongly increase the computational complexity.

Serving the relays with $\tau_i = \log_2(1 + \mathrm{SNR}_i)/R_i$ leads to the effective quantization rate $Q_i = \log_2(1 + \mathrm{SNR}_i)$. That is, relays with high information content (i.e. high SNR_i) get more resources and quantize their observation with a higher rate, as they potentially contribute more to the decoding rate. Furthermore, as the relays with high backhaul rate R_i are considered first (as discussed above) no resources are wasted. Relays with low SNR but high backhaul rates R_i contribute only little information but also do not waste much resources as τ_i gets small, and nodes with low SNR and low backhaul rates R_i are chosen at the end and thus do not harm the strong relays. The quantization noise variance resulting from this choice of τ_i is given as

$$\sigma_{q_i}^2 = \frac{\sigma_{z_i}^2(\lambda_{i,i} + \sigma_{z_i}^2)}{\lambda_{i,i}}, \tag{6.19}$$

which simplifies to $\sigma_{q_i}^2 \approx \sigma_{z_i}^2$ at high SNR. That is, the nodes just get as much resources as necessary to achieve an effective SNR of $\lambda_{i,i}/(\sigma_{q_i}^2 + \sigma_{z_i}^2) \approx 0.5 \cdot \mathrm{SNR}_i$. Once all relays are considered, this leads to

$$\begin{aligned} R^{\mathrm{QF}} &\approx \log_2\left(\det\left(\mathbf{I} + \frac{1}{2}\cdot \mathbf{D}_z^{-1}\mathbf{\Lambda}_{\mathrm{s}}\right)\right) & (6.20)\\ &\approx \log_2\left(\left(\frac{1}{2}\right)^N \cdot \det\left(\mathbf{I} + \mathbf{D}_{\mathrm{z}}^{-1}\mathbf{\Lambda}_{\mathrm{s}}\right)\right) & (6.21)\\ &= \log_2\left(\det\left(\mathbf{I} + \mathbf{D}_{\mathrm{z}}^{-1}\mathbf{\Lambda}_{\mathrm{s}}\right)\right) - N & (6.22) \end{aligned}$$

at high SNR. That is, independent of R_i, there is at least a *quantization noise penalty* of N bps/Hz in the throughput due to the choice of τ_i in Eq. (6.18). However, without iterative optimization, this is a reasonable choice for practical wireless systems, as large gains can be achieved with only limited resources (see Sec. 6.5.1).

Complexity Compared to a gradient search, the implementation of the algorithm can be done very efficiently by computing $\mathbf{D}_{\mathrm{q}}$ once for all $\tau_i = \log_2(1+\mathrm{SNR}_i)/R_i$ and then setting $\mathbf{A} = (\mathbf{D}_{\mathrm{z}} + \mathbf{D}_{\mathrm{q}})^{-1}\mathbf{\Lambda}_{\mathrm{s}}$. In each iteration of adding a new relay, the rows in $\mathbf{A}$ corresponding to non-considered relays can be set to the all-zero-vector (as $\sigma_{q_i}^2 \to \infty$). Hence, computing the inverse is only necessary once in the beginning, whereas it is necessary in each step of the gradient search. Furthermore, at most N iterations of updating the throughput are necessary with the cascade algorithm, leading to a much lower computational complexity, especially for large N.

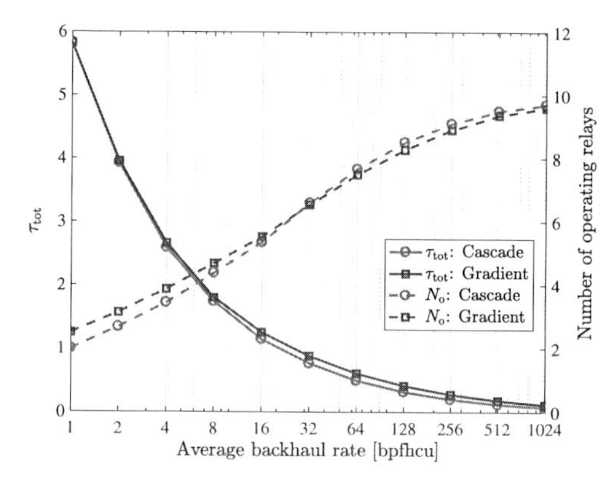

Figure 6.7.: Number of operating relays (dashed lines) and total second hop duration (solid lines) for the cascade resource allocation algorithm and the gradient search in regime III for varying backhaul rates.

6.5.1 Numerical Evaluations

For the evaluation of the cascade algorithm, we consider the same setup as for the relay selection schemes (see Sec. 6.4.1) and compare its performance to a multi-start gradient search with 100 random initializations. However, we only discuss regime III, because the performance characteristics in all regimes are very similar. The performance is evaluated with respect to average backhaul rates $\bar{R} \in \{1, 2, 4, 8, 16, 32, 64, 128, 256, 512, 1024\}$ bpfhcu. That is, the backhaul rates are drawn from $\mathcal{U}(0, 2 \cdot \bar{R})$. Such large backhaul rates could be achieved by using a much larger bandwidth on the second hop relative to the first hop, as discussed in Chap. 8.

At first we investigate the average number of operating relays as shown in Fig. 6.7 with the dashed lines. As to be expected, it increases with increasing average backhaul rate, as the τ_i become shorter and more relays can be served before τ_{tot} decreases the throughput R_{MS} in Eq. (6.5) again. For very low backhaul rates, it is not worth to apply virtual MIMO and only one relay forwards its observations, and for very high backhaul rates nearly all relays are operating all the time. Thereby, the average number of operating relays of the cascade algorithm and the gradient search are very similar. At low backhaul rates, the τ_i of the cascade algorithm in Eq. (6.18) tend to be too large, leading to a lower average number of operating relays. At high backhaul rates, the τ_i are rather too small, leading to a higher average number of operating relays than

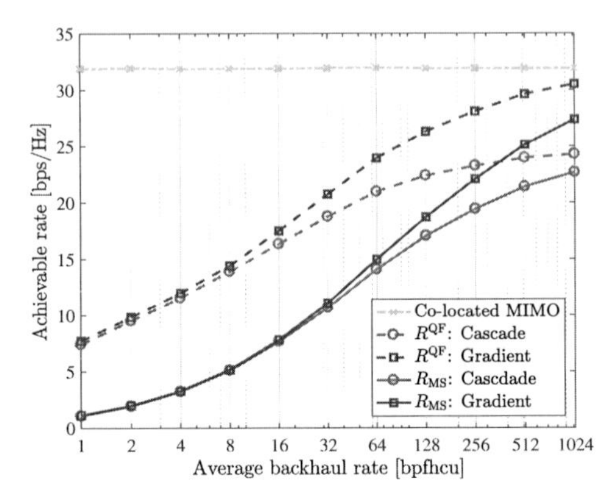

Figure 6.8.: Achievable decoding rate R^{QF} (dashed lines) and achievable throughput R_{MS} (solid lines) for the cascade resource allocation algorithm and the gradient search in regime III for varying backhaul rates.

for the gradient search.

The resulting τ_{tot} is obviously strongly decreasing for increasing backhaul rates, as shown in Fig. 6.7 with the solid lines. That is, while more relays forward their quantized observation, the delay of the virtual MIMO protocol is strongly decreased. For all backhaul rates, the cascade algorithm leads to slightly lower τ_{tot}.

Fig. 6.8 shows the average achievable decoding rate (dashed lines) and the average resulting throughput (solid lines) as well as the average achievable co-located MIMO rate. Considering the achievable decoding rate R^{QF} for the gradient search, it can be seen that at low average backhaul rate $\bar{R}$ a large loss results compared to the theoretically achievable co-located MIMO rate. This is due to the low number of operating relays. For increasing $\bar{R}$, the achievable decoding rate R^{QF} strongly increases. More strong relays become operating and contribute their quantized observation to the decoding until it starts to saturate and slowly achieves the co-located MIMO rate at very high backhaul rates. The additional relays can only contribute little additional information (note that the figure is in logarithmic scale on the x-axis). While the performance of the cascade algorithm closely follows the gradient search for low average backhaul rates $\bar{R}$, the quantization noise penalty as discussed in Eq. (6.22) is clearly visible at high $\bar{R}$. The cascade algorithm can not achieve the co-located MIMO rate, no matter how big the backhaul rates get. Note that the quantization noise penalty is smaller than the predicted N bps/Hz. This is due to the fact that not all relays have a high receive SNR.

For the resulting achievable throughput R_{MS} a similar pattern can be observed. For low average backhaul rates $\bar{R}$, the gap to the co-located MIMO rate is even increasing due to the large τ_{tot}. In this regime, the benefit of the virtual MIMO scheme is limited or not existent at all (see the number of operating relays in Fig. 6.7). For increasing average backhaul rates, the performance of virtual MIMO strongly increases, and for very high average backhaul rates $\bar{R}$ it approaches the achievable decoding rate R^{QF} due to the very low τ_{tot}. As the resulting τ_{tot} is smaller for the cascade algorithm (see Fig. 6.7), the loss compared to the gradient search can be slightly decreased.

Hence, the cascade algorithm is an efficient resource allocation strategy at low to intermediate backhaul rates. At very high backhaul rates larger τ_i than in Eq. (6.18) would be reasonable. However, this would significantly increase the complexity of the algorithm. Furthermore, in practical systems unlikely high backhaul rates are necessary to make a big difference, as will be shown in Chap. 8.

6.6 Conclusions

To optimize the throughput in a quantize-and-forward receive cooperation setup with a shared backhaul, it is important to only consider an optimized subset of the relays forwarding their observation to the final destination. Furthermore, their channel access time and thus the total second hop duration are crucial. The size of the subset as well as the total second hop duration thereby strongly depend on the total number of nodes, the achievable SNRs at the relays and the backhaul rates to the final destination, and have to be adapted consistently. For quasi-stationary systems, where the setup changes slowly, simple relay selection schemes lead to very promising performance, getting close to a computationally expensive gradient search. For non-stationary systems however, a more sophisticated resource allocation strategy is necessary. With the cascade algorithm a promising scheme is presented. It inherently determines the relays in the subset, their channel access time and thus also the total second hop duration and leads to promising results at low computational complexity.

7

Cooperative Communication in Military Mobile Ad Hoc Networks

In the previous chapters, we discussed cooperative communication schemes for mobile ad hoc networks (MANETs) exploiting the diversity gain, the array gain and the spatial multiplexing gain of multi-antenna systems. All schemes have been considered separately and were evaluated in simple setups to focus on their individual potential. In the following two chapters, we evaluate the proposed schemes and combinations thereof in the setup of a military MANET and an urban traffic hotspot of a cellular network.

The focus of this chapter is set on the benefit of cooperative communication schemes in military MANETs considering the transmission range and the spatial multiplexing. In this context, we first compare the distance which can be overcome within a certain number of hops for multi hop transmission, cooperative broadcasting and beam shaping. Afterwards, we are going to demonstrate the potential of virtual multiple-input multiple-output (MIMO) transmit cooperation.

The evaluations are done considering the channel model without small scale fading discussed in Chap. 3, expanded with parameters from real world measurements in military networks. This simple model allows to efficiently evaluate the schemes for a large number of nodes on a conceptual basis. While such evaluations don't allow for detailed statements and comparisons, they provide rough insights into the performance and could indicate suitable operating regimes of the schemes. Thus, they may serve as a decision-making tool in order to set the focus of more detailed evaluations/network design steps.

The remainder of this chapter is structured as follows. In Sec. 7.1 we evaluate and compare the transmission range of multi hop transmission, cooperative broadcasting and beam shaping. In Sec. 7.2 the performance of virtual MIMO is compared to conventional schemes and suitable regimes identified. Sec. 7.3 finally concludes the chapter.

7.1 Range Extension

In the following, we are going to compare the performance of multi hop transmission, multistage cooperative broadcast and beam shaping in a military MANET. To this end, we first introduce a system model which is motivated by a military setup and measurement data. We then evaluate the performance of the mentioned schemes in terms of the distance which can be overcome for various node densities.

7.1.1 System Model

The performance of multi hop transmission and multistage cooperative broadcast are evaluated in a MANET where the locations of the nodes are points of a homogeneous Poisson point process (PPP) Φ with intensity δ in $\mathbb{R}^2$, analogously to the system setup presented in Chap. 3. Without loss of generality, the source is considered to be at the origin. Obviously, military MANETs are not infinitely large and the nodes are not uniformly distributed over the entire network in reality. Nevertheless, this model allows to grasp a reasonable impression of the performance of the considered schemes. As figure of merit we investigate the distance D which can be overcome within a certain number of hops. For the beam shaping we consider a platoon of soldiers (or other nodes), all within one hop communication range of each other, and determine the distance which can be overcome. This setup is sketched in Fig. 7.1.

The channel between any two nodes is determined by a distance dependent path loss according to [86] (Chap. 2.5) with path loss coefficient γ and a random phase shift. Considering Gaussian transmit symbols $s \sim \mathcal{CN}(0, P_{\mathrm{Tx}})$ with transmit power P_{Tx} and additive white Gaussian noise $w \sim \mathcal{CN}(0, \sigma_w^2)$, the received signal of a node at distance d from the source can be written as

$$y = \sqrt{K} \cdot \left(\frac{d}{d_0}\right)^{-\gamma/2} \cdot h \cdot s + w, \tag{7.1}$$

with $h = e^{j\theta}$, $\theta \sim \mathcal{U}(0, 2\pi)$ and d_0 the reference distance for the antenna far field. K denotes a unit-less constant which depends on the antenna characteristics, the wavelength, and the average channel attenuation ([86], Chap. 2.5). For the beam shaping

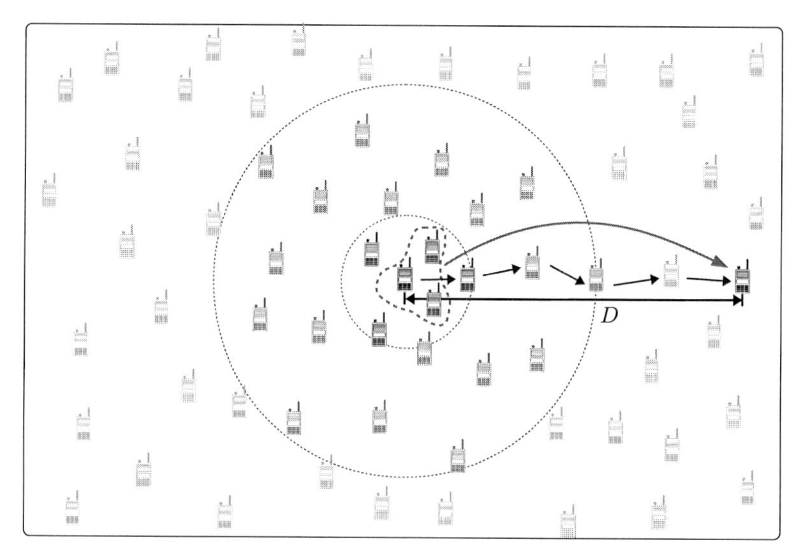

Figure 7.1.: Setup for the performance comparison indicating multi hop transmission, multistage cooperative broadcast and transmit cooperation with beam shaping.

the random phase is neglected, as the radiation pattern is determined considering free space propagation.

Motivated by practical military networks, we consider fixed transmission rates and fixed time slots. No spatial reuse is considered in the same MANET (only one message can be transmitted at a time) and all transmissions are done in-band. That is, we can not resort to a different frequency band for the "short range" data exchange of beam shaping in this scenario. Such out-of-band exchange with bandwidth scaling to boost the performance of beam shaping is considered in 7.2.

The parameters are chosen as considered reasonable in a military network [87]:

- For the transmission we consider a carrier frequency of $f_c = 100$ MHz in the very high frequency (VHF) band, 25 kHz bandwidth and a transmission rate of $C = 10$ kbit/s, which is assumed to be achievable if the signal-to-noise ratio (SNR) at the receiver is at least $\mathrm{SNR}_{\min} = 7$ dB. These numbers are based on measurement data with existing military equipment [87].

- The variance of the thermal noise is determined from

$$\sigma_w^2 = 10 \cdot \log_{10}\left(\frac{4k_B T \Delta f}{0.001}\right) + 8 \quad [\text{dBm}] \tag{7.2}$$

with the Bolzmann constant k_B and a noise figure at the receiver of NF= 8 dBm. Considering a temperature of $T = 300$ K and $\Delta f = 25$ kHz, we get $\sigma_w^2 = 2.62 \cdot 10^{-15}$ W.

- The transmit power per node is set to $P_{\text{Tx}} = 1$ W.

- The path loss coefficient is chosen to be $\gamma = 3.5$. Note that for a scenario considered for beam shaping (e.g. transmitting from a hill), the path loss coefficient is usually smaller, leading to larger gains for the beam shaping than presented in the following.

- Considering $d_0 = 10$ m, the wave length $\lambda = 3$ m (transmit frequency 100 MHz) and setting K to the free space path loss at d_0, $K = (\lambda/(4\pi d_0))^2$ ([86], Chap. 2.5), we get a maximal distance per hop of

$$d_{\text{max}} = \left(\frac{P_{\text{Tx}} \cdot K \cdot d_0^\gamma}{\text{SNR}_{\text{min}} \cdot \sigma_w^2}\right)^{1/\gamma} = 10954 \text{ m} \tag{7.3}$$

for a single node transmitting.

- For the node density δ, two different values are considered: A low density of $\delta = 0.05$ nodes/km^2 and a high density of $\delta = 0.5$ nodes/km^2. This corresponds to an expected number of roughly 19 respectively 188 nodes in the one hop communication range of a single node.

- For the beam shaping, platoon sizes of $N \in \{5, 10, 20, 40\}$ are considered.

7.1.2 Performance Evaluation

The three different communication schemes are evaluated as follows.

Multi hop transmission The multi hop transmission is evaluated as described in Sec. 3.5 for fixed transmission rates considering the maximal communication range $d_{\text{max}} = 10954$ m as given in Eq. (7.3).

Multistage cooperative broadcast The multistage cooperative broadcast is evaluated as described in Sec. 3.4. Analogously to the multi hop transmission, the coverage distances have to be adapted to the modified path loss model in Eq. (7.1), leading to

$$d_{\max,1} = 10954 \text{ m}. \tag{7.4}$$

The minimal energy $P_{\min}$ required for decoding in the further hops (see Eq. (3.20)) is given as

$$P_{\min} = P_{\mathrm{Tx}} \cdot K \cdot d_0^{\gamma} \sum_{n=1}^{N_{k-1}} \mathsf{E}[d_n]^{-\gamma} \approx \mathrm{SNR}_{\min} \cdot \sigma_w^2. \tag{7.5}$$

Beam shaping In transmit cooperation with beam shaping, the source first shares its message with the cooperating nodes in an exchange phase, requiring n_{EX} time slots. Afterwards, the message is jointly transmitted to the destination using another time slot. Due to the large coverage range in a single hop and the assumption that we apply beam shaping from a platoon of soldiers, we consider all nodes to be reachable in one hop in this evaluation, i.e. $n_{\mathrm{EX}} = 1$. Nevertheless, in other setups with much lower coverage range (e.g., in a wireless sensor network) n_{EX} can also be chosen higher to reach enough nodes for the cooperation.

As for leakage based beam shaping (see Chap. 5) the achievable gain in the desired direction strongly depends on the parameters, the node topology, and the desired leakage suppression, we consider the range maximizing solution ($\vartheta = 0$) for the evaluation. Hence, no node topology or parameters have to be specified for the evaluation. That is, no leakage is suppressed and the maximal gain in the desired direction is always achieved. Considering leakage suppression towards hostile units, the gains in the desired direction would slightly decrease as shown in Chap. 5. For a total transmit power of $P_{\mathrm{tot}} = P_{\mathrm{Tx}} \cdot N$, this leads to a total gain in the desired direction of N^2 compared to a single transmitting antenna at P_{Tx}. Hence, if the far-field assumption for the radiation pattern is fulfilled (i.e. if the separation between the cooperating nodes is small compared to D) the received signal power is given as

$$P_{\mathrm{s}} = P_{\mathrm{Tx}} \cdot N^2 \cdot K \cdot d_0^{\gamma} \cdot D^{-\gamma}. \tag{7.6}$$

Via $\mathrm{SNR}_{\min} = P_{\mathrm{s}}/\sigma_w^2$ we can find the distance which can be overcome as

$$D = \left(\frac{P_{\mathrm{Tx}} \cdot N^2 \cdot K \cdot d_0^{\gamma}}{\mathrm{SNR}_{\min} \cdot \sigma_w^2} \right)^{1/\gamma}. \tag{7.7}$$

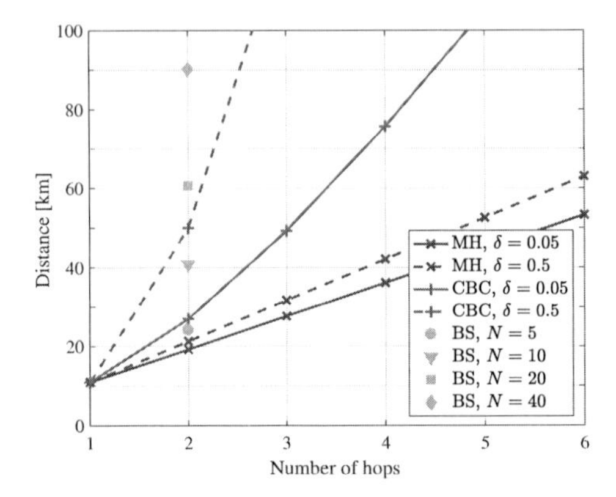

Figure 7.2.: Reachable distance for beam shaping (BS), multistage cooperative broadcast (CBC) and multi hop transmission (MH).

7.1.3 Results

Fig. 7.2 shows the distance which can be reached for multi hop communication, multistage cooperative broadcast and beam shaping. It can be seen that the multistage cooperative broadcast leads to a very fast spreading of the message through the network. That is, a large number of nodes can be reached in only a few hops. Hence, e.g., the overhead to establish a proactive routing protocol can be decreased drastically as shown for military networks in [29] (discussed in more detail in Chap. 2). However, multistage cooperative broadcast is not only a very efficient way to broadcast information to a large number of nodes, but also a reasonable alternative to multi hop transmission for the one to one communication over a large distance. Big gains in the transmission range can be achieved compared to multi hop transmission. These gains can be increased even further at larger node densities as visible from Fig. 7.2. At the same time, the transmission range of multi hop transmission is limited by the number of hops times d_{max}, no matter how high the node density is. Furthermore, no routing is required to reach the destination in multi stage cooperative broadcast. However, on the down side, a large number of nodes is involved, requiring an increased coordination and synchronization and also requiring high transmit power.

Considering beam shaping, it can be seen that large distances can be covered with a reasonable cluster size in only two hops. Compared to multistage cooperative broadcast at larger node density $\delta = 0.5$ nodes/km^2, roughly $N = 20$ nodes are necessary to reach a larger distance. However, cooperative broadcasting involves 188 transmitting nodes in this stage, i.e. much higher transmit power is required. For increasing number of cooperating nodes N, beam shaping clearly outperforms the multistage cooperative broadcast. For a smaller path loss coefficient γ, the gains would be even higher. This can be seen as follows. If we hypothetically consider N nodes placed all in one point, transmitting to a destination with R_{min}, the distance which can be overcome with a transmit diversity scheme as considered in the multistage cooperative broadcast is given as $D = ((P_{\text{Tx}} \cdot N \cdot K \cdot d_0^{\gamma})/(\text{SNR}_{\text{min}} \cdot \sigma_w^2))^{1/\gamma}$. Compared to the distance which can be overcome with beam shaping as denoted in Eq. (7.7), it can be seen that there is a gain in the coverage distance for beam shaping of $N^{1/\gamma}$. This gain is increasing for decreasing γ.

The gain of cooperative broadcast and beam shaping compared to multi hop transmission is clearly visible. To overcome the same distance, much less hops, and thus much less time is necessary. Hence, the throughput of the network can be strongly increased. Of course, the effect scales with the distance to overcome. That is, for smaller distances, the gain is smaller or even inexistent, for larger distances it is larger.

It is important to note that the gains of multistage cooperative broadcast and beam shaping are only achieved for distances over 20000 m (as already the single hop range of multi hop transmission is 10945 m). That is, if we consider the spatial distribution of a standard military unit to be a circle of radius 10000 m (already a spatially large unit), probably all the communication can be done equally good with multi hop communication. However, if there is strong fading and shadowing (strongly decreasing the single hop communication range), or if we have to transmit to a destination in another unit at large distance, the cooperative communication schemes can provide large gains. Furthermore, as described in Chap. 5, leakage based beam shaping can not only be used for range extension, but also to suppress the signal in the direction of hostile units.

7.2 Spatial Multiplexing

In the following, we are going to evaluate the performance of a transmit virtual MIMO scheme serving multiple users simultaneously in a military setup. We then compare it to the conventional way of serving the users sequentially and identify suitable operating regimes.

7.2.1 System Model

To compare the performance of the transmit virtual MIMO scheme with the conventional scheme, we consider a setup with one source and N destinations, randomly distributed on a circle around the source. The radius of the circle is the one hop communication range of a single node $d_{\max}$. That is, all destinations can be reached by the source in one hop. For the transmit virtual MIMO scheme, the source is supported by $N-1$ nodes in its platoon. These nodes are all considered to be in close vicinity of the source. That is, they can all be reached in one hop for the exchange of the transmit data. The channel between any two nodes is modeled identically to the channel in the first scenario. Hence, the channel matrix between the cluster of cooperating nodes and all destinations can be written as

$$\mathbf{H} = \sqrt{K}\left(\frac{d}{d_0}\right)^{-\gamma/2} \cdot \begin{pmatrix} h_{1,1} & \cdots & h_{1,N} \\ \vdots & \ddots & \vdots \\ h_{N,1} & \cdots & h_{N,N} \end{pmatrix}, \tag{7.8}$$

with $h_{i,j} = e^{j\theta}$, $\theta \sim \mathcal{U}(0, 2\pi)$, the random phase from transmitting node j to destination i. All other parameters of the system model are also chosen to be identical to the first scenario. That is, we consider fixed time slots and a fixed transmission rate $R_{\min}$ to the destination (which requires $\mathrm{SNR}_{\min}$). The noise variance is given as $\sigma_w^2 = 2.62 \cdot 10^{-15}$ W, the transmit power is set to 1 W and the path loss coefficient to $\gamma = 3.5$, resulting in a one hop coverage distance $d_{\max} = 10954$ m. However, in contrast to the previous setup, we consider out of band exchange of the transmit data to improve the performance. That is, while the long range transmission to the destinations is done in the VHF band, the local data exchange among the nodes in the transmit cluster is done in the ultra high frequency (UHF) band, where a large amount of bandwidth is available. For reasons of simplicity, we consider a fixed transmission rate $R_{\mathrm{EX}} = \xi \cdot R_{\min}$ for the data exchange with all collaborating nodes. This allows to identify the suitable operating regimes.

7.2.2 Performance Evaluation

The performance evaluation of the two schemes is done with respect to the delay to serve all destinations once, the outage probability and the final throughput.

Conventional transmission To serve the N destinations with R_{min}, N time slots are necessary. This results in a throughput of $R_{\text{MS}} = R_{\text{min}}$ with $P_{\text{out}} = 0$.

Transmit virtual MIMO For the multi-user MIMO precoding, all transmitting nodes need to know all transmit data wich is transmitted with R_{min} to the destinations. If this data is shared with $R_{\text{EX}} = \xi \cdot R_{\text{min}}$, this takes $t_{\text{EX}} = N/\xi$ time slots. As the joint transmission of the precoded signal takes another time slot, the total delay of the protocol is given as $t_{\text{tot}} = 1 + N/\xi$. For the multi-user MIMO precoding, we consider stream wise SLNR precoding [18] with a per stream power constraint of P_{Tx}. That is, for each transmitted stream $n \in \{1, \ldots, N\}$, we determine the precoding vector $\mathbf{w}_n$ by maximizing the signal-to-leakage-plus-noise ratio and scale it such that $\mathbf{w}_n^{\mathsf{H}}\mathbf{w}_n = P_{\text{Tx}}$. The entire precoding matrix is then given by

$$\mathbf{W} = [\mathbf{w}_1, \ldots, \mathbf{w}_N], \tag{7.9}$$

and the transmit vector over the N transmitters as

$$\mathbf{x} = \mathbf{W}\mathbf{s}, \tag{7.10}$$

with the transmit symbol vector $\mathbf{s}$ with i.i.d. elements $\sim \mathcal{CN}(0, 1)$. At the destinations the intended signal is then decoded in the presence of the co-channel interference of the other streams and the thermal noise. If SNR_{min} is not achieved, the destination is considered to be in an outage, resulting in the outage probability P_{out}. The total throughput is then given as $R_{\text{MS}} = (1 - P_{\text{out}}) \cdot N \cdot R_{\text{min}}/t_{\text{tot}}$. To determine the performance of the virtual MIMO schemes Monte Carlo simulations are performed.

7.2.3 Results

In Fig. 7.3 the resulting delay of the two schemes are shown for various exchange rates $R_{\text{EX}} = \xi \cdot R_{\text{min}}$ and number of destinations N. While for $\xi = 1$, the conventional scheme is obviously in advantage, virtual MIMO is paying off for larger ξ. A strong decrease in the delay can be achieved, especially for large virtual MIMO setups.

Nevertheless, while for the conventional scheme each node is achieving R_{min} (as we do not consider fading), this is not guaranteed for virtual MIMO. Due to constructive or destructive addition of the individual signal contributions of the cooperating nodes, the achievable rate at the destinations might be higher or lower than R_{min}. This results

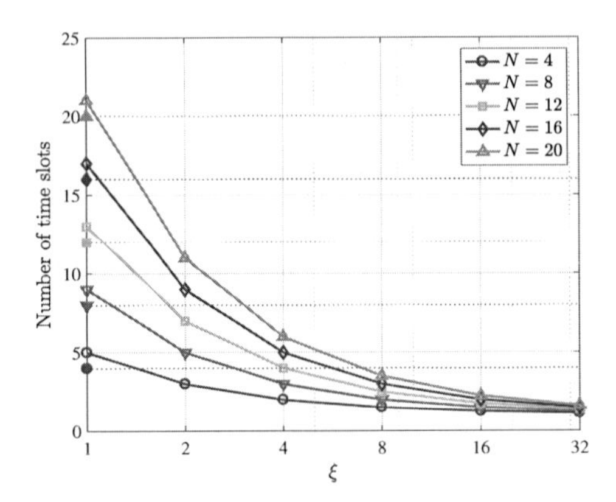

Figure 7.3.: Number of required time slots to serve all destinations as a function of $R_{\text{EX}} = \xi \cdot R_{\text{min}}$. The empty markers and solid lines indicate the virtual MIMO scheme, the filled markers and the dotted lines indicated the conventional transmission.

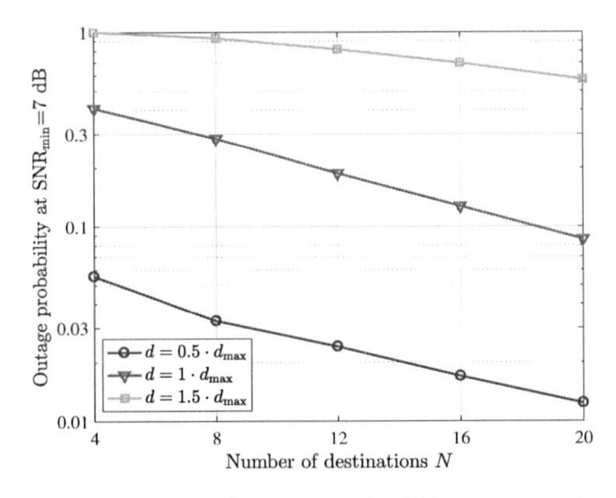

Figure 7.4.: Outage probability at destinations for different source-destination distances, if all nodes are served with R_{min}.

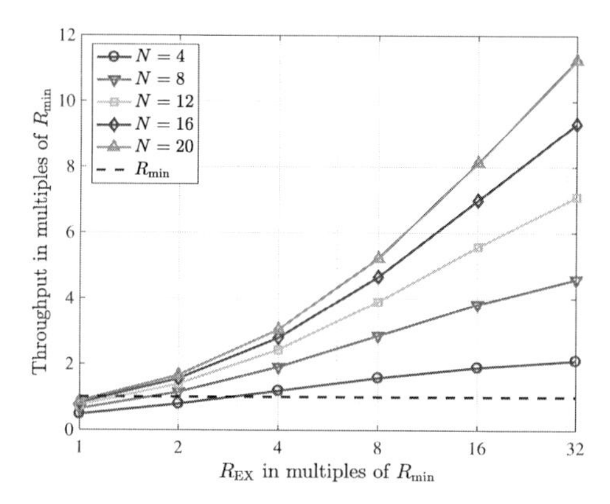

Figure 7.5.: Final throughput of the virtual MIMO scheme compared to sequentially serving the destinations.

in a relatively large outage probability P_{out} especially at low N, as can be seen in Fig. 7.4. This figure also shows the outage probability if the distance between the source and the destination changed. Naturally, the outage probability can be strongly decreased by decreasing this distance, as the receive SNR grows. Note: A simple stream-wise SLNR precoding was applied. By applying a target rate precoding as introduced in Chap. 4, the outage probability could be reduced. However, this was not investigated in the scope of this evaluation.

Considering the achievable throughput, it can be seen in Fig. 7.5 that in contrast to the delay, the benefit of virtual MIMO kicks in later, attenuated by the outage probability. For small number of transmitters and destinations $N = 4$, at least $R_{\mathrm{EX}} = 4 \cdot R_{\min}$ have to be available to outperform the conventional transmission, and at $R_{\mathrm{EX}} = 32 \cdot R_{\min}$ only a factor of 2 in throughput is achieved. However, for larger virtual MIMO systems, large gains can be achieved, emphasizing the huge benefit which cooperative communication can provide in a MANET.

7.3 Conclusions

In this chapter, we compared cooperative communication schemes with conventional schemes in the setup of a military MANET. Considering a simple channel model without small scale fading but based on real world measurements in military MANETs, first rough insights into the performance could be gained. Large potential gains of cooperative communications were demonstrated, and suitable operating regimes were identified.

While most of the transmissions in a military network (inside the own unit) can be efficiently done with a multi hop transmission (usually not more than two hops required), larger distances (e.g., the transmission between two spatially separated units or the transmission to the head quarters) can be efficiently overcome by beam shaping with already a small number of nodes. Similar gains in transmission range can be achieved with multistage cooperative broadcast. Within a few hops a large number of nodes can be reached. Hence, cooperative broadcast is not only an efficient form of distributing information to a large number of nodes, but can also efficiently serve a single destination over large distances, without the necessity of a sophisticated routing protocol. However, the whole network is involved, strongly increasing the energy consumption.

Considering the virtual MIMO scheme, a large number of nodes can be served simultaneously in military MANETs. Using the UHF band where much larger bandwidth is available for the data exchange than in the VHF band, huge gains can be achieved compared to conventional schemes.

8

User Cooperation Enabled Traffic Offloading in Urban Hotspots

In this chapter, we combine three of the previously discussed cooperative communication schemes for mobile ad hoc networks (MANETs) in order to serve the mobile stations in an urban traffic hotspot with ultra high user density.

Serving mobile stations (MSs) in an area with ultra high user density, such as a busy public square or a sports stadium, is a challenging problem. Various proposals have been made for 5G, the next generation of wireless networks, such as massive multiple-input multiple-output (MIMO) systems [30], millimeter wave (mmW) communication [31], or network densification with heterogeneous networks and traffic offloading to wireless local area networks (WLANs) [32]. However, the performance of massive MIMO is limited by the potentially correlated scattering in such a dense environment [32]. In mmW communication, the high number of RF chains necessary for the large antenna arrays is a limiting factor, as they are very power consuming [88]. Furthermore, for both approaches, huge investments into infrastructure (more antennas/ access points) would be necessary to serve the large amount of users. Traffic offloading to existing WLAN access points (further called residential backhaul access points (RBAPs)) scales poorly in the number of mobile stations, as the mobile stations have to be served sequentially in time by a single RBAP (or by a few RBAPs if the hotspot size allows for spatial reuse).

In this context, we propose to serve the mobile stations by *user cooperation enabled traffic offloading*. That is, by forming a virtual antenna array, multiple mobile stations jointly offload their traffic to multiple typically privately owned WLAN access points. This way, multiple mobile stations can be separated by distributed spatial multiplexing without additional infrastructure.

The proposed user cooperation enabled traffic offloading is a combination of transmit multi-user MIMO with leakage based precoding (see Chap. 4) and quantize-and-forward virtual MIMO receive cooperation (see Chap. 6) for the long haul RBAP access, and cooperative broadcast (see Chap. 2) for the local exchange. As shown in the previous chapter, a cluster of cooperating nodes can efficiently serve multiple other nodes simultaneously if the local exchange can be implemented efficiently, i.e. if a large bandwidth is available compared to the long haul RBAP access. To this end, we propose the local exchange to be performed in the 60 GHz band where a large amount of bandwidth is available. The RBAP access is performed in the 2.4 GHz band. In contrast to the rather conceptual evaluations in Chap. 7 which should serve as rough performance estimates, we consider a sophisticated numerical simulation framework with realistic parameters and channel models in this chapter, and provide a rigorous feasibility study of user cooperation enabled traffic offloading in an urban hotspot scenario. We thereby demonstrate large gains compared to conventional schemes by exploiting the large available bandwidth in the 60 GHz band for the local exchange. Thereby, cooperative broadcasting is a key enabler for an efficient data exchange at 60 GHz as it can efficiently circumvent the large path loss and the strong shadowing.

This chapter is mainly based on our work published in [36] and [38]. The remainder is structured as follows. In Sec. 8.1 related work is reviewed and in Sec. 8.2 the system setup is discussed. Sec. 8.3 introduces the user cooperation protocol and Sec. 8.4 covers the simulation framework. In Sec. 8.5, Sec. 8.6 and Sec. 8.7 the performance of the protocol is evaluated considering the uplink, the downlink and a spatial reuse scheme. Sec. 8.8 finally concludes the chapter.

8.1 Related Work

Virtual MIMO schemes have been repeatedly proposed in literature for cellular networks, e.g., in [22] or [89]. In [22], distributed-MIMO multi-stage relaying networks are investigated with respect to the resource allocation. As a specific subset of such networks, mobile stations in a cellular network are proposed to form a virtual antenna array for capacity enhancement. In [89], conferencing mobile stations on orthogonal channels are investigated to enhance multi-cell processing for the uplink of cellular networks based on a Wyner model.

Different to the mentioned approaches, we consider traffic offloading to RBAPs and (i) take into consideration local users (LUs), which are assigned to the individual RBAPs, and the mutual interference between the local users and offloaded traffic and

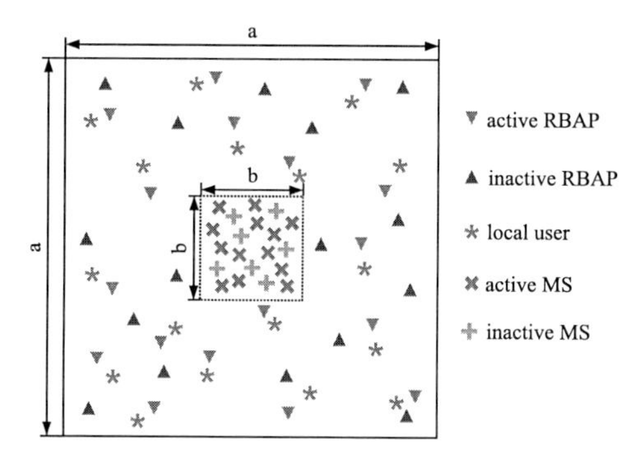

Figure 8.1.: Urban hotspot scenario.

(ii) require that all signal processing shall be done on the mobile station side, such that the RBAPs do not have to distinguish between offloaded traffic and local users. This allows distributed ownership of the RBAPs as no cooperation among them is necessary, and makes the protocol fully transparent for the RBAPs. Hence, the protocol can be overlaid on the existing infrastructure, without any central processing. These constraints mandate a careful design of both, the local exchange and the access phase, which are typical for virtual MIMO. We propose a specific protocol implementation, provide a rigorous feasibility study and identify suitable operating regimes.

As an extension to the urban hotspot scenario described in this chapter, we also applied user cooperation enabled traffic offloading in a remote hotspot scenario in [37]. By a remote hotspot we refer to an area with ultra high user density located outside but close to a densely populated area with many RBAPs. This could, e.g., be a music festival, a sports event or an emergency situation, where a large amount of people needs to get help. In such areas cellular networks normally do not support big crowds of people. Hence, traffic offloading to the RBAPs in the close vicinity is a reasonable temporary approach. As the protocol is basically the same, however, supplemented by a local user fairness scheme, this setup is not further discussed in this thesis and the interested reader is referred to [37].

8.2 System Setup

The setup of consideration is shown in Fig. 8.1. It represents an urban traffic hotspot with ultra high user density, surrounded by buildings with many RBAPs. The traffic hotspot, modeled by a square of side length b, contains N_{MS} users which want to transmit/receive data (active MSs) and $\bar{N}_{\mathrm{MS}}$ users which do not want to communicate (inactive MSs). Nevertheless, the inactive users can potentially disturb the data exchange among the active mobile stations by shadowing the line-of-sight between active users. In the area around the traffic hotspot we consider a larger number of RBAPs. We again distinguish between N_{AP} active RBAPs and $\bar{N}_{\mathrm{AP}}$ inactive RBAPs. While the inactive RBAPs are in idle mode, each active RBAP is communicating with a local user (LU) in its vicinity. Hence, there are $N_{\mathrm{LU}} = N_{\mathrm{AP}}$ local users. All nodes are considered to be equipped with a single omnidirectional antenna and to be half-duplex.

8.3 User Cooperation for Ultra Dense Environments

In order to serve the mobile stations in the hotspot, they form a virtual antenna array and jointly communicate with the RBAPs in the surrounding. Hence, the traffic of the hotspot is distributed over a large area without additional infrastructure. To support distributed ownership, the wireless system has to be designed such that it minimizes the coordination between the involved RBAPs. Thus, all signal processing is done on the mobile station side, and the RBAPs are accessed by distributed spatial multiplexing. This makes the protocol fully transparent for the RBAPs and the hotspot simply appears like a virtual local user to them.

In the uplink, the mobile stations first exchange their transmit data and then jointly transmit to the RBAPs. That is, they are all considered to be synchronized (in time, frequency and phase). To separate the streams for the RBAPs sophisticated multi-user MIMO precoding is necessary. In the downlink, the non-cooperating RBAPs simultaneously but independently transmit to the cooperating mobile stations in the hotspot. In order to achieve spatial multiplexing and to decode the messages, the mobile stations then quantize their analog received signals and broadcast them among each other. Eventually, all mobile stations can decode the messages and extract the data intended for them. The uplink and downlink are described in more detail and are evaluated in Sec. 8.5 respectively Sec. 8.6. In the following, we are going to discuss the RBAP assignment, the medium access control, and the protocol implementation for the feasibility study.

8.3.1 RBAP Assignment and Medium Access Control

For the RBAP access, the mobile stations can incorporate both active and inactive RBAPs, e.g., based on their channel strength. If an active RBAP is accessed, the corresponding local user is turned off during the access phase to mitigate interference, but is still served during the exchange phase. Hence, the transmissions of the hotspot and the local users have to be coordinated carefully.

In the optimal case, the virtual MIMO scheme can be simply overlaid on the existing infrastructure and protocols without any further coordination. To this end, a deterministic approach can be applied. That is, all RBAPs are accessed individually by the hotspot for the protocol setup by applying carrier sense multiple access/collision avoidance (CSMA/CA). If a collision with local users occurs, the procedure is simply repeated until the RBAP can be accessed. The hotspot then transmits a time-table to each of them which defines fixed time slots for the local users and the hotspot. This time-table is forwarded from the RBAPs to all assigned local users. Once all involved nodes know the time-table, the protocol can be applied without further coordination.

Setting up the protocol, synchronizing the involved users (in time, frequency and phase) and coordinating their transmission, as well as the acquisition of the required channel state information are crucial elements of the proposed scheme to work efficiently. Nevertheless, we do not further focus on these parts in this thesis, but rather investigate the performance once the protocol is set up.

8.3.2 Protocol Implementation

Not all active mobile stations in the cluster have to be served simultaneously. To decrease the coordination among the mobile stations and potentially also to improve the performance, sub-clusters of M of the N_{MS} active mobile stations are considered. To allow for M independent streams simultaneously M RBAPs are assigned to such a sub-cluster. Multiple sub-clusters can then be served sequentially or also simultaneously with a certain spatial reuse pattern.

The proposed protocol consists of two phases: The long-haul MIMO access phase (AC), where the mobile stations communicate with the RBAPs, and the local exchange phase (EX), where the mobile stations exchange their transmit or receive data. For the performance evaluation we consider the implementation of these phases as follows.

The long-haul MIMO access phase The RBAP access phase is considered to be done in the 2.4 GHz industrial, scientific and medical (ISM) band in order to be compatible with the standard WLAN protocols.

The RBAPs for the AC phase are assigned according to their channel strength. That is, considering the row vector $\bar{\mathbf{h}}_i$ to be the channel vector from the cooperating mobile stations to RBAP i, we choose the M RBAPs with the highest $p_i = \bar{\mathbf{h}}_i \bar{\mathbf{h}}_i^{\mathsf{H}}$.

For the spatial multiplexing in the downlink, we consider quantize-and-forward virtual MIMO with the cascade resource allocation algorithm as discussed in Chap. 6. To achieve the spatial multiplexing in the uplink, we apply stream-wise SLNR precoding according to [18] (see Sec. 4.3.3) and consider a per-stream power constraint P_{MS}. That is, for each transmitted stream $m \in \{1, \ldots, M\}$ we determine the precoding vector $\mathbf{w}_m$ by maximizing the signal-to-leakage-plus-noise ratio and scale it such that $\mathbf{w}_m^{\mathsf{H}} \mathbf{w}_m = P_{\mathsf{MS}}$. The entire precoding matrix is then given by

$$\mathbf{W} = [\mathbf{w}_1, \ldots, \mathbf{w}_M]. \tag{8.1}$$

For the leakage term of a single stream all RBAPs which are accessed either by the hotspot (except the desired RBAP) or by local users are considered.

The local exchange phase In order to allow for the large gains of virtual MIMO schemes as demonstrated in Chap. 6 and Chap. 7, the local exchange is considered to be done in the 60 GHz license free band where large bandwidths are available.

To distribute the information in the exchange phase, dynamic cooperative broadcasting is considered (see Chap. 2). That is, whenever a node is able to decode the message, it immediately starts to retransmit it using a different codebook. As discussed in Chap. 2, this assumption is somewhat idealistic, as all mobile stations would need to know the codebooks used by all other involved mobile stations and the time when they start to transmit. However, in contrast to multistage cooperative broadcast, it allows to evaluate the performance of the protocol without defining the minimal exchange rate R_{min} (which would strongly impact the performance and thus would need to be optimized). It can thus be considered as an upper bound on the performance with multistage cooperative broadcast.

The broadcasting rate of a node is then determined by the achievable rate with which the weakest of all involved mobile stations can be reached. However, by applying cooperative broadcasting, we can circumvent the problem of the high path loss and the strong shadowing at 60 GHz with a single omnidirectional antenna. In a hotspot with ultra high user density the nodes are very close. Hence, as the message is flooded through the clusters from node to node, only small distances have to be overcome for which high rates can be achieved. That is, the weak channels can be bypassed by relays, and the message can be distributed very efficiently to a large number of mobile stations without routing or beam forming.

As a reference for the cooperative broadcasting, a simple broadcast scheme is considered as well. Thereby, the originating mobile station simply transmits its signal without support of other mobile stations, until all could decode the message. Hence, the performance is limited by the weakest channel to all involved mobile stations.

8.4 Simulation Framework

The performance evaluation is done with numerical simulations in a setup as sketched in Fig. 8.1. The width of the setup is set to $a = 600$ m and the width of the hotspot to $b = 50$ m. $N_{\mathrm{AP}} = 140$ active and $\bar{N}_{\mathrm{AP}} = 140$ inactive RBAPs are randomly distributed in the respective area, whereby a minimal distance of $d_{\mathrm{min,AP}} = 40$ m between all active RBAPs and between all inactive RBAPs has to be fulfilled. This minimal distance should reflect the necessary separation between simultaneously working RBAPs due to CSMA/CA. The local users are randomly placed around their assigned RBAPs within a distance of $10 \leq d_{\mathrm{min,LU}} \leq 20$ m. The active and inactive users are randomly placed within the hotspot area with a minimal distance of 1 m to each other. Variable numbers of active and inactive users are evaluated, but they are always chosen to be equal, $N_{\mathrm{MS}} = \bar{N}_{\mathrm{MS}}$. The minimal distance between any RBAP and mobile stations in the hotspot is set to 10 m.

All channels are considered frequency flat fading. For the 60 GHz channel between the mobile stations in the hotspot, the path loss and shadowing is drawn from the log-distance path loss model of [31], which is based on real world measurements in New York. This path loss model is summarized in App. A.3. For each link we determine whether it is line-of-sight or not. To this end, each user in the hotspot is modeled as a circle with a diameter of 0.6 m. Whenever the connection between the centers of two users is blocked by another user, the link is considered non-line-of-sight, otherwise line-of-sight. For the line-of-sight channels Ricean fading is assumed with a K factor which is log-normal distributed with mean $\mu = 7$ and standard deviation $\sigma = 3$, accounting for the strong line-of-sight link [69]. For the non-line-of-sight channels Rayleigh fading is assumed. All other channels in the network are considered to be non-line-of-sight channels with path loss and shadowing coefficients drawn according to the WINNER II scenario C2 channel model [69]. This channel model is summarized in App. A.2. All antenna heights are set to 1.5 m, and the transmit frequency is set to 2.4 GHz. For the mobile stations in the hotspot, block shadowing to the RBAPs is assumed. That is, the hotspot is split up into equally distributed squares of 10 m width. All mobile stations in one square then observe the same shadowing to a specific RBAP. In all simulations, we assume that the required channel state information is available at all mobile stations. That is, $\bar{\mathbf{h}}_m^{\mathsf{H}}\bar{\mathbf{h}}_m$ and $\mathbf{F}_m^{\mathsf{H}}\mathbf{F}_m$ for each stream $m \in \{1,\dots,M\}$, with $\bar{\mathbf{h}}_m$ the channel vector from the cooperating mobile stations to the assigned RBAP for stream m, and $\mathbf{F}_m$ the channel from the cooperating mobile stations to all RBAPs considered in the leakage term for stream m.

The transmit power of the local users is set to $P_{\mathrm{LU}} = 0.01$ W. The transmit power of the mobile stations in the hotspot is set to $P_{\mathrm{MS}} = 0.1$ W in the access phase and to $P_{\mathrm{EX}} = 0.1$ W and $P_{\mathrm{EX}} = 0.01$ W for comparison in the exchange phase. The RBAPs transmit with $P_{\mathrm{LU}} = 0.01$ W to the local users and with $P_{\mathrm{MS}} = 0.1$ W to the mobile stations. The bandwidth at 2.4 GHz is assumed to be 20 MHz and the corresponding receiver noise variance is set to $\sigma_w^2 = 10^{-12}$ W. The bandwidth for the exchange at 60 GHz is set to $\beta \cdot 20$ MHz, with the bandwidth scaling factor β. The corresponding noise variance is then $\beta \cdot \sigma_w^2$ W.

For reasons of simplicity, we consider the local users and the mobile stations to be synchronized in their uplink and downlink in all simulations. As the local users have the same transmit power as the RBAPs transmitting to the local users, the interference level for the mobile stations in the hotspot is not expected to change significantly if they are not synchronized in their up- and downlink.

8.5 Uplink Performance Evaluation

In the uplink of the protocol, spatial multiplexing is achieved by sophisticated distributed multi-user MIMO precoding. This procedure is illustrated in Fig. 8.2, considering a single sub-cluster with M participating mobile stations. In order to be able to compute the uplink signal individually at each mobile station, all transmit data needs to be available at all mobile stations. To this end, the mobile stations share their messages u_m, $m \in \{1,\dots,M\}$, in the exchange phase. These messages are then mapped onto the transmit symbols s_m, $m \in \{1,\dots,M\}$, intended for the assigned RBAPs. In

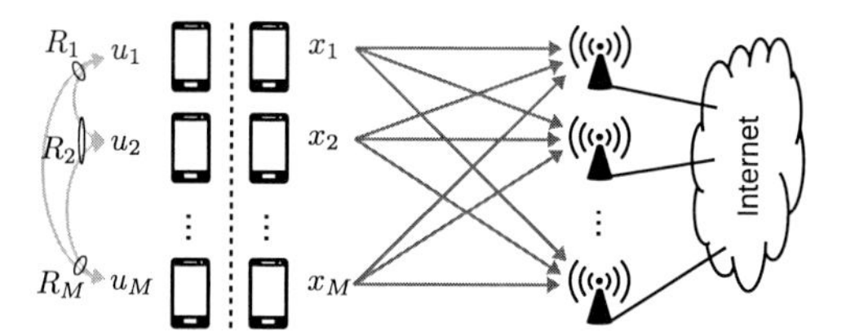

Figure 8.2.: Illustration of the virtual MIMO transmit cooperation.

the access phase, the mobile stations then transmit a superposition of the transmit symbols given by the elements of $\mathbf{x} = \mathbf{W}\mathbf{s}_{\text{MS}}$, with $\mathbf{W} \in \mathbb{C}^{M \times M}$ the precoding matrix for the spatial multiplexing and $\mathbf{s}_{\text{MS}} = [s_1, \ldots, s_M]^\mathsf{T}$ the transmit signal vector of the mobile stations in the sub-cluster. The received signal of RBAP i during the access phase can then be written as

$$y_i = \bar{\mathbf{h}}_i \mathbf{W}\mathbf{s}_{\text{MS}} + \sum_{j \in \mathcal{J}} f_{i,j} s_{\text{LU},j} + w_i, \tag{8.2}$$

where $\bar{\mathbf{h}}_i \in \mathbb{C}^{1 \times M}$ is the channel vector from the mobile stations in the sub-cluster to RBAP i, $\mathcal{J}$ is the set of currently transmitting local users, $f_{i,j}$ the channel from local user j to RBAP i, $s_{\text{LU},j}$ the transmit symbol of local user j and $w_i \sim \mathcal{CN}(0, \sigma_w^2)$ is additive white Gaussian noise. The transmit symbols of the mobile stations are considered i.i.d. $\sim \mathcal{CN}(0, P_{\text{MS}})$ and the symbols of the local users i.i.d. $\sim \mathcal{CN}(0, P_{\text{LU}})$. Each RBAP then decodes its intended signal in the presence of the thermal noise, the interference of the local users and the co-channel interference from the streams of the hotspot intended for other RBAPs. The achievable sum rate in the access phase of the mobile stations in the sub-cluster is then given as the sum of the individually achievable decoding rates of all involved RBAPs and is denoted by $R_{\text{MS}}^{\text{AC}}$.

To achieve maximal fairness, each mobile station shares $R_{\text{MS}}^{\text{AC}}/M$ bits per channel use of the access phase with all other involved mobile stations[1]. The time it takes until this exchange is completed for all mobile stations is denoted by τ_{tot}. For the clarity of exposition, we, without loss of generality, normalize all time intervals to the duration of 1 channel use of the access phase in the following. Hence, combining the exchange and the access phase, the final achievable sum rate of the mobile stations is then given

[1]Alternatively, each mobile station could be explicitly assigned to one RBAP. Hence, an RBAP would only receive the data of one specific mobile station. The amount of data a mobile station has to share would then be proportional to the achievable rate at its RBAP.

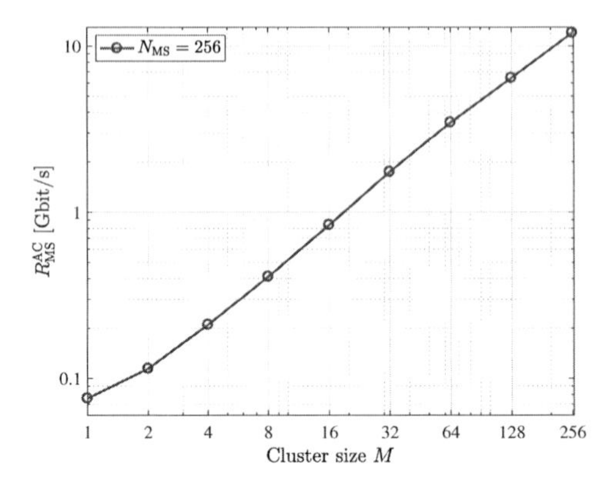

Figure 8.3.: Performance in the access phase for fixed number of mobile stations ($N_{MS} = 256$) and varying cluster size considering local users assigned to the RBAPs. Lower number of mobile stations in the hotspot lead to almost the same curves, as the performance does not strongly depend on the node density.

as

$$R_{MS} = \frac{R_{MS}^{AC}}{1 + \tau_{tot}}. \tag{8.3}$$

8.5.1 Performance Results and Discussion

The performance evaluation is done considering the achievable sum rate in the access phase, the duration of the exchange phase relative to the access phase (τ_{tot}) and the resulting achievable sum rate of the mobile stations averaged over 1000 Monte Carlo simulations. In the first simulations, a single cluster with M out of the N_{MS} mobile stations is considered, i.e. only a subset of the mobile stations cooperates at a time. The clusters are formed by picking the first mobile station randomly and then assigning the $M - 1$ nearest neighbors to it. This setup is evaluated for $\beta = 10$ and $N_{MS} \in \{64, 128, 256\}$ with clusters sizes $M \in \{1, 2, 4, 8, 16, 32, 64, 128, 256\}$ (up to the maximal possible number $M \leq N_{MS}$).

Fig. 8.3 shows the average achievable sum rate in the access phase for $N_{MS} = 256$. All other considered N_{MS} are not shown, as the curves are almost identical. As expected, the achievable sum rate in the access phase increases roughly linearly with increasing number of nodes in the cluster, as more RBAPs can be accessed at once.

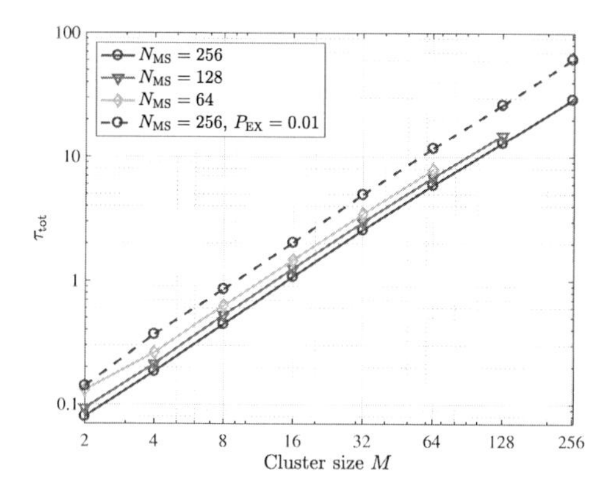

Figure 8.4.: Performance in the exchange phase considering fixed number of mobile stations and varying cluster size, for transmit power 0.1 W (solid lines) and 0.01 W (dashed lines) for comparison.

In Fig. 8.4 the corresponding τ_{tot} are shown, the duration of the exchange phase relative to the access phase. For a transmit power of $P_{\text{EX}} = 0.1$ W for the mobile stations in the exchange phase two observations can be made. The first is that for larger sub-cluster size M, the relative duration of the exchange phase τ_{tot} is increasing. This is mainly due to the higher sum rates which can be achieved in the access phase. Thus more data needs to be exchanged, taking a longer time. Furthermore, for a larger sub-cluster size M the spatial dimension of the sub-cluster is bigger. Hence, higher distances have to be overcome, leading to lower efficiency. The second observation which can be made is that τ_{tot} is smaller for the same sub-cluster size M at higher N_{MS}, although the same amount of data has to be exchanged. As the spatial dimension of a sub-cluster with M mobile stations is smaller at higher node density, the efficiency in the exchange is increased due to the smaller distance to be overcome. Also shown in this figure is the performance for $P_{\text{EX}} = 0.01$ W with $N_{\text{MS}} = 256$ mobile stations (dashed line), where the exchange efficiency clearly suffers. The lower transmit power decreases the achievable rates between the nodes, leading to lower exchange rates and thus longer τ_{tot}.

Combining the access and the exchange phase, an optimal cluster size can be found in terms of the mobile station sum rate. This is shown in Fig. 8.5. The maximal achievable rate is not achieved at the maximal M. Hence, we do better if we only consider a subset of the nodes in the hotspot at once. Furthermore, already with relatively few

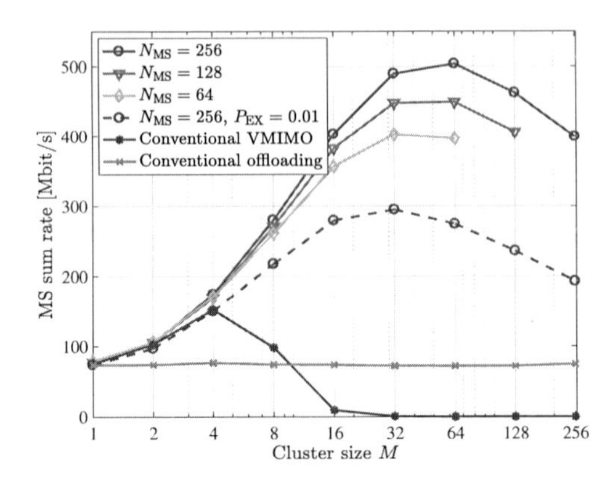

Figure 8.5.: Average achievable mobile station sum rate considering fixed number of mobile stations and varying cluster size, for transmit power 0.1 W and 0.01 W, and conventional offloading as well as conventional virtual MIMO for $N_{MS} = 256$ as reference.

nodes ($M = 16$) very good performance can be achieved. This significantly reduces the complexity, as less channel state information and less coordination is necessary. Fig. 8.5 also shows the performance of conventional broadcasting in the exchange phase for $N_{MS} = 256$ instead of cooperative broadcasting (denoted by conventional VMIMO). As the performance of the conventional broadcast in a cluster depends on the weakest channel, it gets very poor, as the number of nodes per cluster and with it the distances to overcome increases. This impressively shows how efficiently the cooperative broadcast can deal with the problem of the high path loss with a single omnidirectional antenna at 60 GHz. Comparing $P_{EX} = 0.1$ W with $P_{EX} = 0.01$ W, a significant performance drop can be observed, analogously to the performance drop in the exchange phase. However, while the mobile station sum rate decreases, also the energy consumption is decreased. Compared to a conventional traffic offloading approach, where one mobile station after the other accesses an RBAP, large gains can be obtained.

Considering the relative duration of the exchange phase τ_{tot} in Fig. 8.4, it can be seen that the exchange phase is the bottleneck of the protocol for large M (as it takes much longer than the access phase). The efficiency in the exchange can be improved by increasing the bandwidth scaling factor β or by applying a spatial reuse. In Fig. 8.6, we show the achievable rates for different exchange bandwidths, $\beta \in \{1, 10, 20, 30, 40, 50\}$ for $N_{MS} = 256$ and $M = 64$. Increasing the bandwidth strongly decreases the time

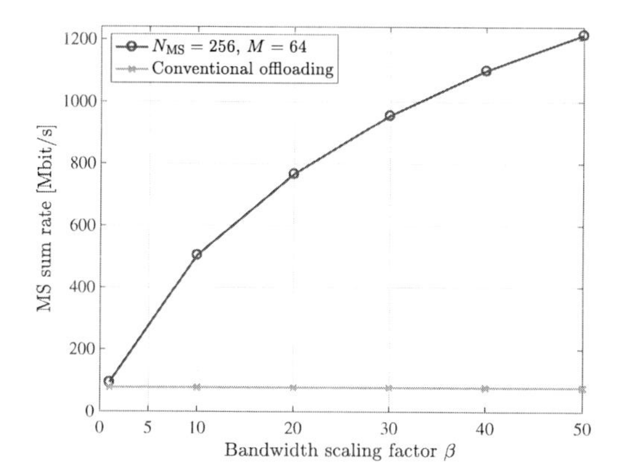

Figure 8.6.: Achievable rates for 256 mobile stations, $M = 64$ and varying β.

needed for the exchange phase, leading to much higher achievable rates. However, the achievable mobile station sum rate can not be increased arbitrarily by simply scaling the bandwidth, as it is limited by the performance in the access phase. Hence, a slow saturation of the achievable mobile station sum rate can be observed in Fig. 8.6. Improving the performance by applying a spatial reuse in the exchange and access phase is discussed in Sec. 8.7.

8.6 Downlink Performance Evaluation

In the downlink of user cooperation enabled traffic offloading the non-cooperating RBAPs simultaneously but independently transmit to the cooperating mobile stations. The mobile stations then quantize and forward their received signal to each other in order to decode the message, as discussed in Chap. 6 for quantize-and-forward virtual MIMO receive cooperation. This procedure is illustrated in Fig. 8.7. The received signal of mobile station i in the downlink access phase is given by

$$y_i = \sum_{m=1}^{M} h_{i,m} s_{\text{MS},m} + \sum_{j \in \mathcal{J}} g_{i,j} s_{\text{LU},j} + w_i, \tag{8.4}$$

with $h_{i,m}$ the channel between RBAP m and mobile station i, $\mathcal{J}$ the set of currently served local users and $g_{i,j}$ the channel from the RBAP serving local user j to mobile

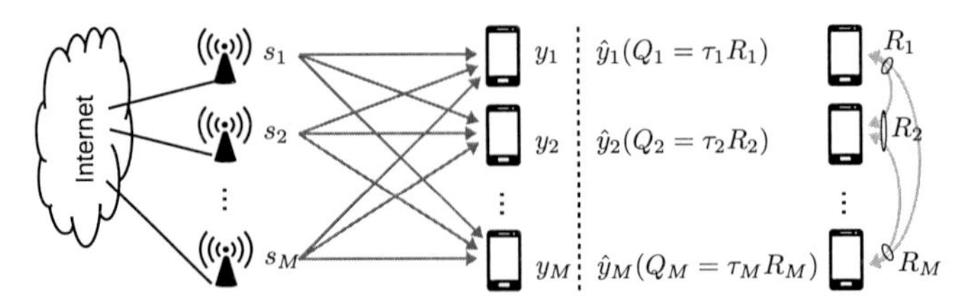

Figure 8.7.: Illustration of the quantize-and-forward virtual MIMO receive cooperation.

station i. The transmit symbols are assumed to be $s_{\text{MS},m} \sim \mathcal{CN}(0, P_{\text{MS}})$ for the mobile stations and $s_{\text{LU},j} \sim \mathcal{CN}(0, P_{\text{LU}})$ for the local users. $w_i \sim \mathcal{CN}(0, \sigma_w^2)$ denotes additive white Gaussian noise. The variance of the received signal is thus given as

$$\sigma_{y_i}^2 = \underbrace{P_{\text{MS}} \sum_{m=1}^{M} |h_{i,m}|^2}_{\lambda_{i,i}} + \underbrace{P_{\text{LU}} \sum_{j \in \mathcal{J}} |g_{i,j}|^2 + \sigma_w^2}_{\sigma_{z_i}^2}, \tag{8.5}$$

with the signal power $\lambda_{i,i}$ and the interference plus noise power $\sigma_{z_i}^2$.

The achievable rate R_i with which each mobile station can forward its quantized observation to the other mobile stations is determined by the broadcasting rate. Considering the channel access time τ_i of mobile station i, its effective quantization rate is then given as $Q_i = \tau_i \cdot R_i$, resulting in zero mean circularly complex quantization noise with variance

$$\sigma_{q_i}^2 = \frac{\sigma_{y_i}^2}{2^{Q_i} - 1} = \frac{\sigma_{y_i}^2}{2^{\tau_i R_i} - 1} \tag{8.6}$$

if vector quantizers are assumed at the mobile stations. Different to the setup in Chap. 6, there is no final destination to which the mobile stations forward their observation. Instead, the mobile stations share their observation with each other. To this end, each mobile station has its own signal always in un-quantized form available, even if it is not forwarding the signal to the other mobile station. Still, we consider the achievable decoding rate to be given by

$$R^{\text{QF}} = \log_2 \det \left(\mathbf{I} + (\mathbf{D}_\text{z} + \mathbf{D}_\text{q})^{-1} \mathbf{\Lambda}_\text{s} \right), \tag{8.7}$$

with $\mathbf{I}$ the identity matrix, the signal covariance matrix $\mathbf{\Lambda}_\text{s} = P_{\text{MS}} \cdot \mathbf{H}\mathbf{H}^{\text{H}}$, where $\mathbf{H} \in \mathbb{C}^{M \times M}$ denotes the channel between the RBAPs and the mobile stations with the elements $h_{i,m}$, $\mathbf{D}_\text{z} = \text{diag}\left(\sigma_{z_1}^2, \ldots, \sigma_{z_M}^2\right)$ the interference plus noise covariance matrix and $\mathbf{D}_{\mathrm{q}} = \mathrm{diag}\left(\sigma_{q_1}^2, \ldots, \sigma_{q_M}^2\right)$ the quantization noise covariance matrix. That is, we consider a lower bound on the achievable sum rate, which is the same for all mobile stations, simplifying the rate allocation for the RBAPs. Hence, if we again normalize all time intervals to the time 1 channel use in the access phase takes, the mobile station sum rate is given as

$$R_{\mathrm{MS}} = \frac{R^{\mathrm{QF}}}{1 + \tau_{\mathrm{tot}}}, \tag{8.8}$$

with $\tau_{\mathrm{tot}} = \sum_{m=1}^{M} \tau_m$ the total duration of the exchange phase relative to the access phase. To optimize the mobile station sum rate, the resources in the exchange phase have to be assigned carefully as discussed in Chap. 6.

8.6.1 Performance Results and Discussion

For the performance evaluation, we consider the cascade resource allocation algorithm introduced in Chap. 6 and compare its performance to a multi-start gradient search. To this end, we consider the averaged results over 1000 Monte Carlo simulations. As the impact of the node density was already investigated for the uplink, we consider a fixed number of active mobile stations $N_{\mathrm{MS}} = 256$ and serve a single sub-cluster of size M at a time. Again, the sub-clusters are formed by picking the first mobile station randomly and then assigning the $M - 1$ nearest neighbors to it. Specifically, we consider the sub-cluster sizes $M \in \{1, 2, 4, 8, 16, 32, 64, 128, 256\}$ and evaluate the performance for $\beta = 10$ and $P_{\mathrm{EX}} = 0.1$ W.

At first, we investigate the average number of operating mobile stations selected to forward their observation and the resulting total duration of the exchange phase relative to the access phase, shown in Fig. 8.8. It can be seen that for both resource allocation schemes, the gap between the number of operating mobile stations and the cluster size M is strongly increasing with increasing cluster size. At large cluster sizes only a small subset of the nodes is forwarding its information. This mainly comes from the fact that for increasing cluster size the achievable broadcasting rates are strongly decreasing. More mobile stations need to be reached and larger distances have to be overcome. Hence, less information can be exchanged in the same time and it is not worth to consider more operating mobile stations to forward their observation as the increase in the decoding rate can not compensate the longer duration of the exchange phase.

The average number of operating mobile stations selected by the cascade algorithm is clearly below the corresponding number for the gradient search for large cluster sizes.

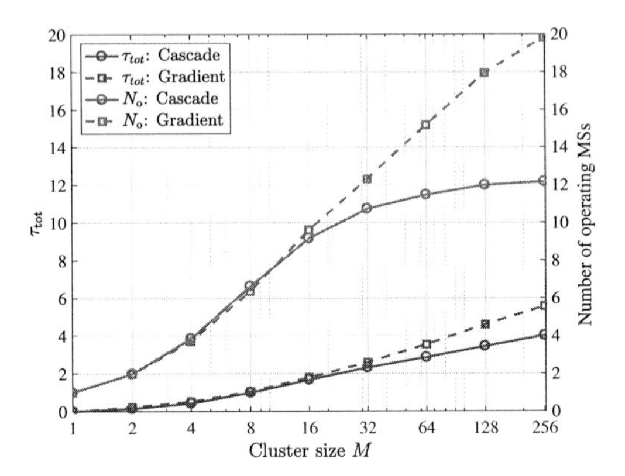

Figure 8.8.: Number of operating mobile stations which forward their observation, and total duration of exchange phase.

This comes from the strict relation of τ_i and R_i in Eq. (6.18). For the decreasing R_i at large M the assigned resources τ_i are potentially too large, leading to a lower number of considered mobile stations. Analogously to the number of operating mobile stations, τ_{tot} only increases slowly with increasing cluster size as can be seen in Fig. 8.8.

The number of operating mobile stations also strongly impacts the average decoding rate R^{QF} as shown in Fig. 8.9. As only a limited number of additional mobile stations forward their quantized observations at large sub-cluster size, the achievable decoding rate starts to saturate. Due to the suboptimal number of operating mobile stations and the channel access time assigned to them, a large gap between the achievable decoding rate of the gradient search and the cascade algorithm can be observed.

Nevertheless, considering the resulting achievable mobile station sum rate R_{MS}, the lower resulting τ_{tot} of the cascade resource allocation partially compensate the loss in the decoding rate R^{QF} as can be seen in Fig. 8.10. Although a mostly suboptimal subset of mobile stations is considered, the performance of the cascade resource allocation is getting close to the gradient search at much lower complexity. As already observed for the uplink in Sec. 8.5, the best performance is achieved if multiple sub-clusters are served sequentially (peak at $M = 32$). That is, a higher throughput can be achieved at lower system complexity (coordination and medium access control).

Fig. 8.10 also shows the performance of the uplink in comparison to the downlink. As can be seen, the downlink outperforms the uplink for a low cluster size M. This

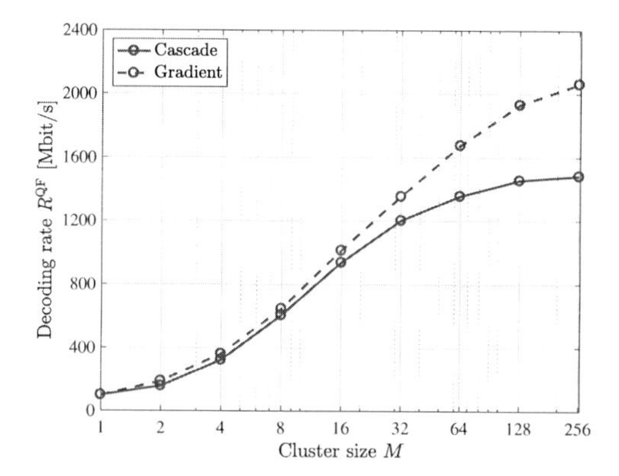

Figure 8.9.: Average achievable decoding rate R^{QF}.

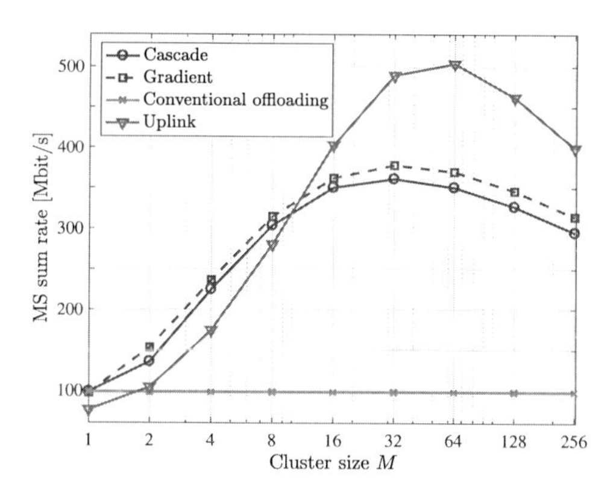

Figure 8.10.: Average system throughput R_{MS}.

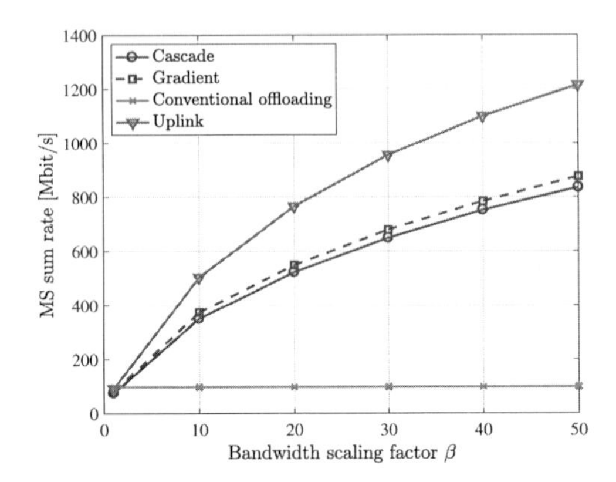

Figure 8.11.: Bandwidth scaling for $M = 64$.

comes from the fact that at low cluster size many local users are still communicating with their RBAPs. As these local users are potentially close to the RBAPs assigned to the hotspot, their interference is strong in the uplink. In the downlink in contrast, the active RBAPs transmitting to their local users are farther away from the mobile stations in the hotspot. Hence, the interference is lower. Furthermore, a large portion of the mobile stations forward their observation (see Fig. 8.8), leading to a high spatial multiplexing gain. At increasing M however, more and more local users are turned off in the uplink, leading to a lower interference at the RBAPs assigned to the hotspot. In this regime, the uplink clearly outperforms the downlink due to the strong impact of the quantization noise and the limited number of mobile stations which forward their information. Still, compared to a conventional offloading scheme, where the mobile stations are served sequentially, huge gains can be achieved in the downlink, already for a limited cluster size.

With increasing bandwidth in the exchange phase higher broadcasting rates can be achieved. This leads to a strongly increased system throughput as shown in Fig. 8.11 for $M = 64$. The cascade algorithm closely follows the performance of the gradient search even for very large bandwidth. That is, for practical operating regimes as shown here, the cascade resource allocation is a reasonable trade-off between performance and computational complexity. As we evaluated the performance for $M = 64$ in this figure, the achievable throughput in the uplink is substantially higher than in the downlink for all bandwidth scaling factors β.

8.7 Spatial Reuse and Local User Performance

Large gains have been shown to be achievable by user cooperation enabled traffic offloading in the previous sections. In this section we consider a spatial reuse scheme in the uplink of the protocol which can further improve the performance. The same scheme can also be applied in the downlink to increase the performance. Similar gains are expected as the amount of interference plus noise roughly scales in the same way in the downlink. However, its evaluation is omitted in this thesis. The increased gains for the mobile stations in the hotspot lead to a decreased performance of the local users. To this end, we also evaluate their performance in this section.

Spatial reuse As we have seen in the previous sections, smaller sub-clusters are more efficient in the information exchange and also require less channel state information. In order to further increase the performance of user cooperation enabled traffic offloading, a spatial reuse can be applied in the exchange as well as in the access phase. That is, the hotspot is split up into multiple small sub-clusters, each operating individually with a certain reuse pattern. As the signal propagation in the 60 GHz band strongly suffers from shadowing and thus only leads to limited mutual interference between the sub-clusters, a high reuse factor can be applied in the exchange phase.

To this end, we introduce the number of sub-clusters N_{C} and the reuse factors for the exchange phase (r^{EX}) and the access phase (r^{AC}), where one out of r^{EX} respectively r^{AC} clusters is active simultaneously. The achievable rate of cluster c in the access phase is denoted by $R^{\mathrm{AC}}_{\mathrm{MS},c}$. For reasons of simplicity, we assume that the duration of the access phase is the same for all clusters in the performance evaluation. Thus, if again all time intervals are normalized to the duration of 1 channel use in the access phase, each mobile station in cluster c shares $R^{\mathrm{AC}}_{\mathrm{MS},c}/M$ bits with each other mobile station in the sub-cluster. The necessary time thereof is denoted by $\tau^{\mathrm{EX}}_{\mathrm{tot},c}$. We assume that all simultaneously active clusters need to be done with their data exchange before the next set of clusters can start an exchange or an access phase. That is, the exchange time of the set of concurrently exchanging clusters $\mathcal{C}_k$ is given by

$$\tau^{\mathrm{EX}}_{\mathrm{max},i} = \max_{c \in \mathcal{C}_k}(\tau^{\mathrm{EX}}_{\mathrm{tot},c}). \tag{8.9}$$

This is illustrated in Fig. 8.12 for one round of the protocol for $N_{\mathrm{C}} = 16$ and $r^{\mathrm{EX}} = r^{\mathrm{AC}} = 4$. The total exchange time of all clusters is then given as

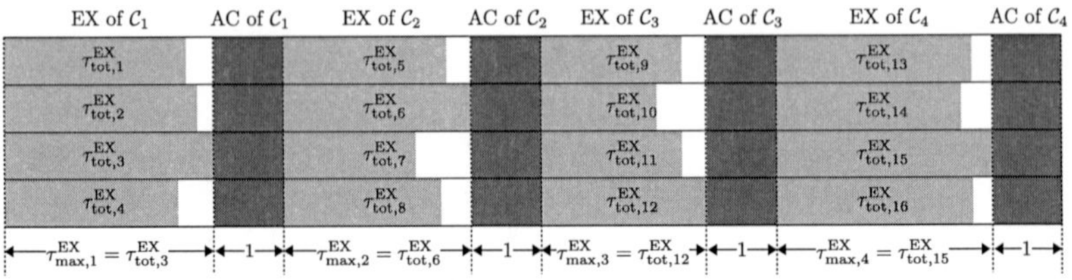

Figure 8.12.: Illustration of one round of the reuse protocol for $N_{\mathrm{C}} = 16$ and $r^{\mathrm{EX}} = r^{\mathrm{AC}} = 4$.

$$\tau_{\mathrm{tot}} = \sum_{i=1}^{r^{\mathrm{EX}}} \tau_{\mathrm{max},i}^{\mathrm{EX}}, \tag{8.10}$$

and the sum of the achievable rates over all clusters as

$$R_{\mathrm{MS,tot}}^{\mathrm{AC}} = \sum_{c=1}^{N_{\mathrm{C}}} R_{\mathrm{MS},c}^{\mathrm{AC}}. \tag{8.11}$$

The achievable mobile station sum rate then follows as

$$R_{\mathrm{MS}} = \frac{R_{\mathrm{MS,tot}}^{\mathrm{AC}}}{r^{\mathrm{AC}} + \tau_{\mathrm{tot}}}. \tag{8.12}$$

Splitting up the hotspot into small clusters and applying a spatial reuse can improve the performance in the exchange phase. However, it also means that less mobile stations are available per cluster for the distributed precoding in the access phase, leading to smaller array and multiplexing gains. Furthermore, the clusters need to coordinate the reuse pattern and their RBAP choices.

Local users Because in the access phase several RBAPs are accessed by the mobile stations in the hotspot, several local users might be unable to communicate with their RBAP. Furthermore, the other local users might suffer from the increased interference level induced by the mobile stations in the hotspot. The achievable sum rate of all local users during the access phase is denoted by $R_{\mathrm{LU}}^{\mathrm{AC}}$. During the exchange phase, all local users can communicate with their RBAPs. Furthermore, no additional interference is generated by the mobile station in the hotspot, as they exchange their data out of band. The achievable rate of the local users during the exchange phase is denoted by $R_{\mathrm{LU}}^{\mathrm{EX}}$. Hence, the achievable rate of the local users averaged over the exchange and

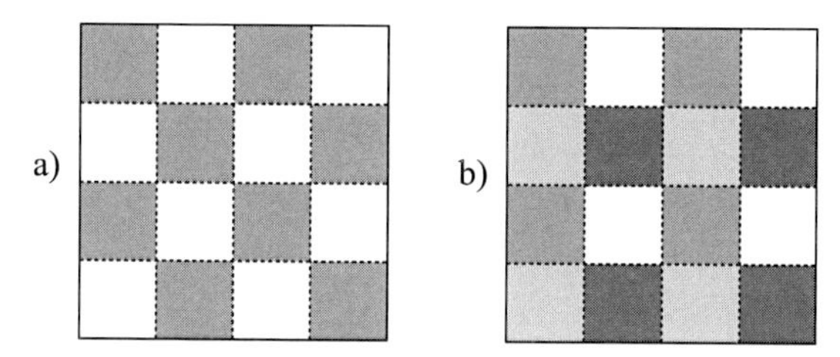

Figure 8.13.: 16 uniformly arranged clusters with: a) reuse pattern for reuse factor 2, b) reuse pattern for reuse factor 4. The clusters with the same shade of gray operate simultaneously.

access phase is then given as

$$R_{\mathrm{LU}} = \frac{R_{\mathrm{LU}}^{\mathrm{AC}} + R_{\mathrm{LU}}^{\mathrm{EX}} \cdot \tau_{\mathrm{tot}}}{1 + \tau_{\mathrm{tot}}}. \tag{8.13}$$

The backhaul access rate, i.e. the sum rate of all RBAPs, averaged over the exchange and access phase can be found as

$$R_{\mathrm{b}} = R_{\mathrm{MS}} + R_{\mathrm{LU}}. \tag{8.14}$$

8.7.1 Performance Results and Discussion

In the following, we investigate the performance if a spatial reuse is applied, averaged over 1000 Monte Carlo simulations. To this end, we consider $N_{\mathrm{C}} = 16$ uniformly placed clusters, with $M \in \{1, 4, 8, 12, 16\}$ randomly placed mobile stations per cluster (i.e. $N_{\mathrm{MS}} = \{16, 64, 128, 256\}$) and different reuse factors $r^{\mathrm{EX}}, r^{\mathrm{AC}} \in \{1, 2, 4, 16\}$, with reuse patters as shown in Fig. 8.13. For reasons of simplicity we consider only $r^{\mathrm{EX}} = r^{\mathrm{AC}}$ in the evaluations. Note that $M = 1$ corresponds to a conventional offloading scheme with spatial reuse (no exchange phase). If a reuse $r^{\mathrm{AC}} < N_{\mathrm{C}}$ is applied, we use a simple fairness algorithm, which assigns the RBAPs to the clusters in a round robin fashion according to their channel strengths (as described in Sec. 8.3.2), until all clusters have their required number of RBAPs.

The performance in the access phase is shown in Fig. 8.14 by considering the sum of the achievable rates over all sub-clusters divided by the total duration of the access phase, i.e. $R_{\mathrm{MS,tot}}^{\mathrm{AC}}/r^{\mathrm{AC}}$. Obviously, the performance is increasing for a higher number of nodes per sub-cluster M, as more streams are transmitted in total. By decreasing

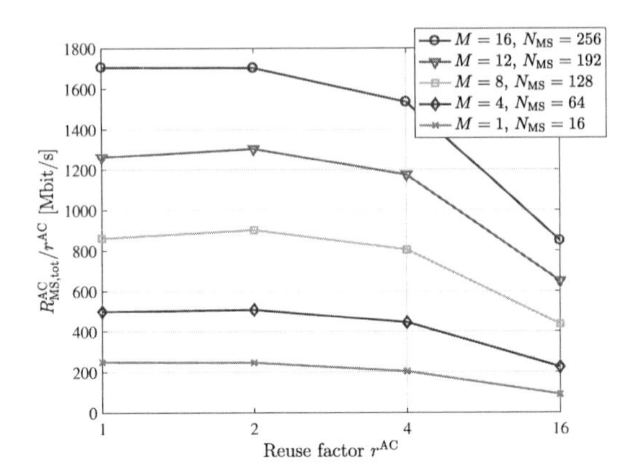

Figure 8.14.: Performance in the access phase considering a spatial reuse: sum of the achievable rates over all sub-clusters divided by the total duration of the access phase.

the reuse factor from 16 down to 2 (i.e. more clusters are active simultaneously), the achievable sum rate is strongly increased, even though the clusters interfere with each other in the RBAP access. That is, the decreased $R^{AC}_{MS,c}$ per sub-cluster due to the interference is more than compensated. For a reuse factor of 1, no further gain can be achieved, as the additional interference between the sub-clusters can not be compensated anymore.

For the evaluation of the exchange phase we consider $\beta = 10$ and $P_{EX} = 0.1$ W. The resulting performance is shown in Fig. 8.15. Note that τ_{tot} reflects the total exchange time which also depends on the amount of data to exchange (see Fig. 8.14). That is, this figure does not directly reflect the efficiency of the exchange. However, by jointly considering Fig. 8.14 and Fig. 8.15, it can be seen that the efficiency in the exchange phase can be strongly increased by increasing the spatial reuse from $r^{EX} = 16$ to $r^{EX} = 4$ (i.e. more clusters are active simultaneously). Due to the spatial separation and the strong shadowing at 60 GHz, there is a strong interference separation between simultaneously active clusters. Hence, more data can be exchanged in less time. Due to the increased amount of data to exchange (see Fig. 8.14) and the slightly decreasing efficiency because of strong interference (not visible in this figure), the total duration of the exchange phase τ_{tot} increases again for $r^{EX} = 2$. For a reuse factor of $r^{EX} = 1$ (i.e. all clusters are exchanging simultaneously), the performance strongly suffers, as the mutual interference is too strong. τ_{tot} increases severely although not more data needs

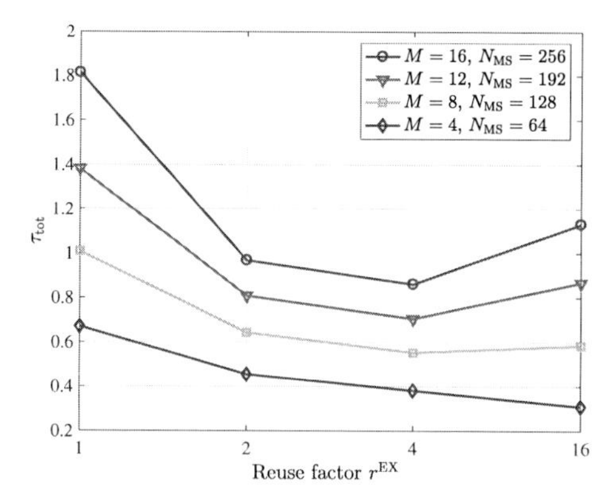

Figure 8.15.: Performance in the exchange phase considering a spatial reuse.

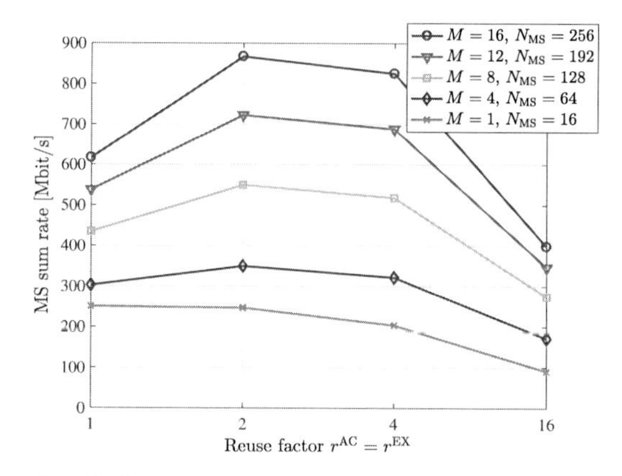

Figure 8.16.: Final achievable mobile station sum rate considering a spatial reuse.

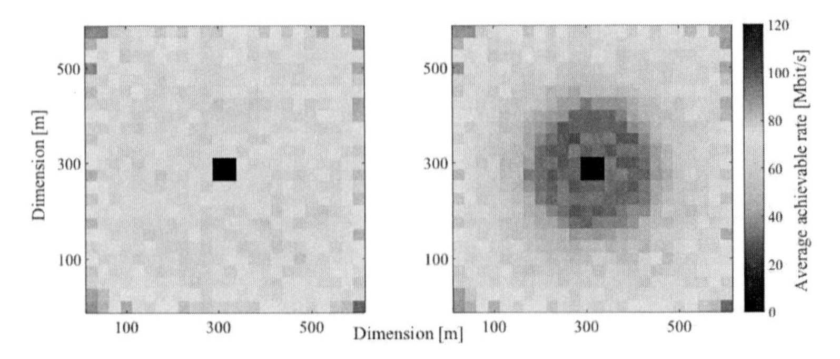

Figure 8.17.: LU rates during EX phase (left) and averaged over EX and AC phase (right) for 16 clusters with 16 nodes each and spatial reuse 2.

to be exchanged. For a sub-cluster size of $M = 1$ no exchange is necessary. Hence, the corresponding curve is missing in this figure.

The resulting mobile station sum rate is shown in Fig. 8.16. The best results can be achieved for a reuse factor of $r^{\mathrm{EX}} = r^{\mathrm{AC}} = 2$. The access as well as the exchange phase become much more efficient due to the spatial reuse and scale beneficially with the number of nodes. Already for $M = 1$ a spatial reuse leads to large gains compared to conventional offloading ($M = 1$, $r^{\mathrm{EX}} = r^{\mathrm{AC}} = 16$). By increasing the cluster size and therefore applying virtual MIMO, the gains can be further increased.

While large gains are achievable for the mobile stations in the hotspot, the local users strongly suffer. This can be seen in Fig. 8.17. On the left side, it shows the spatial distribution of the LU rates during the exchange phase. Similar average local user rates are achieved independent of the location, except for edge effects (reduced interference level at the edge of the setup). The dark blue square in the middle is the hotspot where no RBAPs and local users are located. Comparing these rates to the rates on the right side of Fig. 8.17, where the average local user rates over both phases are shown for $N_{\mathrm{C}} = 16$ with $r^{\mathrm{EX}} = r^{\mathrm{AC}} = 2$ and $M = 16$, it can be seen that the local users around the hotspot suffer strongly due to the frequent RBAP access of the hotspot.

Considering the resulting backhaul rates in Tab. 8.1, it can be seen that they do not drastically change, as the loss of the local users is compensated by the gain of the mobile stations. Only the multiple cluster scheme with spatial reuse 2 notably suffers, as many local users have to be turned off frequently.

Fig. 8.18 in the end shows the achievable rates for different exchange bandwidths,

Scheme	Without hotspot	Conventional offloading	$N_C = 1$ $M = 64$	$N_C = 16$ $r=2$, $M=16$
R_b	9.81	9.71	9.94	8.55

Table 8.1.: Backhaul rates [Gbit/s], i.e. average sum rates of all RBAPs ($N_{MS} = 256$).

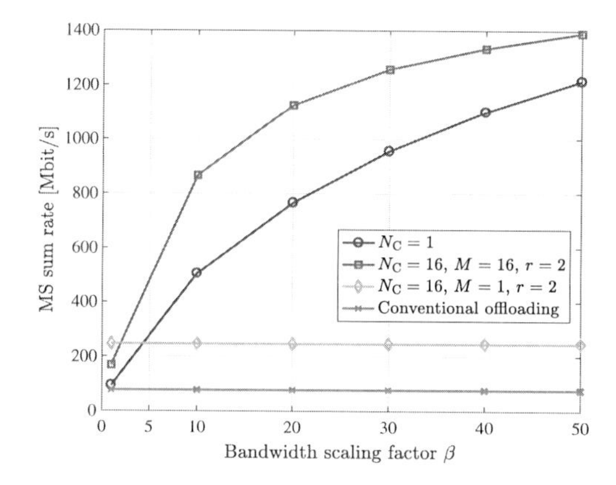

Figure 8.18.: Achievable rates for 256 mobile stations and varying β. $M = 16$ for $N_C = 16$ and $M = 64$ for $N_C = 1$.

$\beta \in \{1, 10, 20, 30, 40, 50\}$. The results are shown for $N_{MS} = 256$ and considering a single cluster with $M = 64$, as well as 16 clusters with $M = 16$ and spatial reuse of $r^{EX} = r^{AC} = 2$. Compared to the achievable gains without a spatial reuse (blue curve), the gains can be further boosted by applying a spatial reuse, especially at intermediate bandwidth scaling factors. For very high bandwidth scaling factors, the gains compared to no spatial reuse diminish. This is due to the very low τ_{tot} which can be achieved. Hence, the access phase is becoming the bottleneck of the protocol and is dominating the performance. As the access phase is more efficient for large cluster sizes and without a spatial reuse, the benefit of a spatial reuse is decreased.

8.8 Conclusions

In this chapter, we proposed a transparent user cooperation protocol based on a combination of cooperative communication schemes for MANETs. It allows to efficiently serve the mobile stations in an urban traffic hotspot by traffic offloading to the WLAN access points in the surrounding. We discussed the necessary coordination among the mobile stations and the local users assigned to the access points, and proposed a system implementation with user cooperation at 60 GHz with cooperative broadcast, and WLAN access at 2.4 GHz with SLNR precoding in the uplink respectively quantize-and-forward virtual MIMO with cascade resource allocation in the downlink. With a numerical simulation framework considering realistic channel models and local users assigned to the WLAN access points, it has been shown that 60 GHz communication combined with cooperative broadcasting is very efficient for the user cooperation, even with a single omnidirectional antenna per node. The large available bandwidth in the 60 GHz band allows for an efficient exchange, while the problem of high path loss and strong shadowing is circumvented by cooperative broadcasting. Furthermore, the high path loss and strong shadowing is even beneficial if a spatial reuse in the hotspot is considered, as it provides a strong interference separation among the sub-clusters which operate simultaneously.

Considering only 16 out of 256 mobile stations at a time and 10 times more bandwidth in the exchange phase than in the access phase, a 5 fold increase in the mobile station sum rate compared to conventional offloading can be achieved. Due to the low sub-cluster size, this requires only limited channel state information and coordination among the nodes. By splitting up the hotspot into 16 clusters with 16 nodes each and applying a spatial reuse in 1 out of 2 clusters, even a gain of roughly 10 compared to conventional offloading can be achieved. Compared to 16 clusters with only 1 mobile station per cluster, a more than 3 fold increase in the throughput can be achieved, underlining the potential of the user cooperation scheme. While the performance of the conventional offloading does not scale with the number of users, the achievable sum rate can be increased for the virtual MIMO scheme by increasing the number of nodes per cluster. By scaling the bandwidth for the user cooperation up to 50 times the bandwidth for the access phase, the gain can be even further increased. Hence, it is a promising approach to serve the mobile stations in an urban traffic hotspot.

9 Summary and Conclusion

In this thesis we investigated low complexity physical layer cooperative communication schemes for mobile wireless networks to improve the coverage, the throughput and the scalability. Specifically, we focused on the following three schemes:

- For multistage cooperative broadcast, we derived a performance prediction based on the order statistics of the distances of the cooperating nodes to a destination outside of the cooperation cluster. It allows to accurately approximate the performance for large networks at low computational complexity and thus simplifies the network design and operation. At the same time, the derived order statistics can be applied to predict the coverage distance of multi hop transmission with high accuracy.
- With leakage based beam shaping we introduced a low complexity pattern synthesis approach for distributed beam shaping in mobile ad hoc networks (MANETs). With a closed-form solution, large gains in the desired direction can be achieved while efficiently suppressing the signal in undesired directions, although the radiation patterns become very irregular and spiky due to the random node arrangement and the possibly large separation between the nodes.
- Considering a quantize-and-forward virtual multiple-input multiple-output (MIMO) receive cooperation scheme to allow multiple sources to communicate simultaneously with one single antenna destination, we proposed low complexity resource allocation schemes for the shared backhaul to the destination. They allow to efficiently optimize the system throughput and lead to promising results.

We provided theoretical analysis as well as numerical evaluations of the proposed schemes and identified suitable operating regimes in two scenarios: Military MANETs and urban hotspots with ultra high user density.

In the urban traffic hotspot scenario, we combined user cooperation with traffic offloading. With a hybrid protocol implementation with offloading in the 2.4 GHz band

and local exchange among the cooperating nodes in the 60 GHz band, large gains are demonstrated compared to non-cooperating approaches. Thereby, cooperative broadcast is shown to be a key enabler for the large gains. Due to its cooperative diversity, it can efficiently circumvent the large path loss and strong shadowing in the 60 GHz band, even with a single omnidirectional antenna per node. Thus, it allows for very high rates in the local exchange. This strongly boosts the performance and leads to large gains compared to conventional approaches. By increasing the exchange bandwidth and applying a spatial reuse, even larger gains can be achieved.

For any transmit virtual MIMO scheme a sophisticated precoding is necessary to achieve spatial multiplexing. In this context, we investigated the relation between transmit power and leakage power in leakage based precoding, a promising multi-user MIMO precoding approach. As the performance of mobile stations in a multi-user MIMO scheme are inherently coupled via the generated leakage power, the precoding has to be designed carefully in order to optimize the system performance. Based on the gained insights, we proposed a decentralized target rate precoding and a rate optimal precoding under a joint transmit power and leakage power constraint. They allow to optimize the system performance in terms of throughput and outage probability and can efficiently be applied in cellular setups with coordinated multipoint transmission as well as in MANETs. For both schemes, a quasi closed-form solution was presented.

In summary, the investigated physical layer cooperative communication schemes can strongly increase the performance in military MANETs and urban traffic hotspots. Large gains in terms of transmission range, throughput and scalability can be achieved at low complexity.

Appendix

A.1 Directive Antenna Pattern

In the following, we are going to describe the antenna pattern used in the cellular simulation setup for the numerical evaluations in Chap. 4. With the permission of the author, this section is adopted from [90].

The considered directive antennas have a pattern as recommended by 3GPP [91]:

$$A(\theta) = G - \min\left\{12\left(\frac{\theta}{\theta_{3\mathrm{dB}}}\right)^2, A_{\mathrm{m}}\right\} \text{ [dB]}, \quad -180 \leq \theta \leq 180, \tag{A.1}$$

with

- θ being the angle between the direction of interest and the boresight of the antenna (angle between the main lobe of the antenna array and the mobile of interest),
- $A_{\mathrm{m}} = 25$ dB the maximal attenuation,
- $\theta_{3\mathrm{dB}} = 65°$ the 3 dB (half-power) beamwidth, and
- $G = 17$ dBi, the antenna gain with respect to an isotropic antenna element.

The antenna pattern is depicted in Fig. A.1

A.2 WINNER II, Scenario C2 Channel Model

The scenario C2 of the WINNER II channel model [69] corresponds to an urban micro-cell environment with a high density of buildings. In the non-line-of-sight propagation

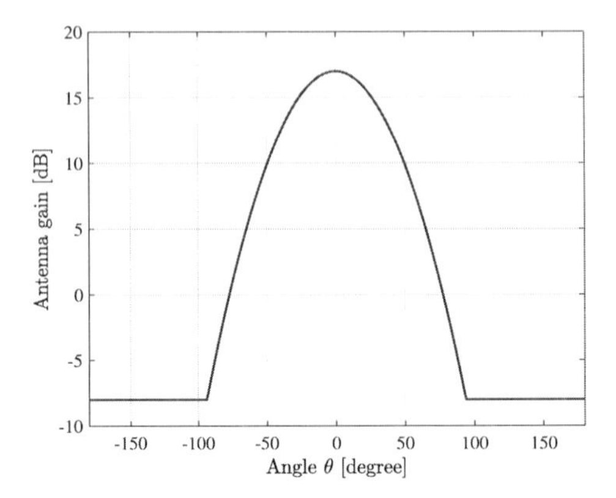

Figure A.1.: Antenna pattern as considered in the cellular network setup.

conditions, the channels are modeled as Rayleigh fading with a path loss and shadowing component given as

$$L = \left(44.9 - 6.55\log_{10}\left(\frac{h_{\mathrm{BS}}}{\mathrm{m}}\right)\right)\log_{10}\left(\frac{d}{\mathrm{m}}\right) + 34.46 \\ + 5.83\log_{10}\left(\frac{h_{\mathrm{BS}}}{\mathrm{m}}\right) + 23\log_{10}\left(\frac{f_c}{5\mathrm{GHz}}\right) \text{ [dB]}, \quad \text{(A.2)}$$

with h_{BS} the height of the base station antennas, d the distance between the transmitter and the receiver and f_c the carrier frequency

As we do not consider line-of-sight scenarios, the corresponding fading, shadowing and path loss coefficients are omitted here.

A.3 60 GHz Log-Distance Path Loss Model

The path loss and shadowing parameters we use for the 60 GHz channels in Chap. 8 are drawn according to [92] which is based on the log-distance path loss model in [31] derived from measurements in New York city at 73 GHz. It is given as

$$L = 20\log_{10}\left(\frac{4\pi d_0}{\lambda}\right) + 10\gamma\log_{10}\left(\frac{d}{d_0}\right) + X_\sigma \text{ [dB]}, \quad \text{(A.3)}$$

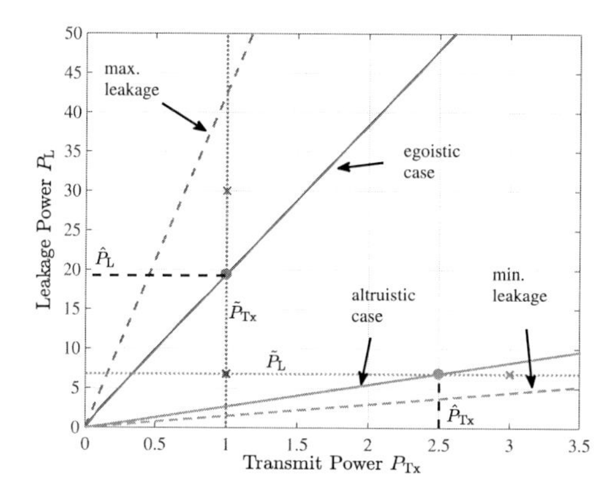

Figure A.2.: The transmit power - leakage power plane.

with d_0 the reference distance (1 m), λ the wavelength, γ the path loss coefficient, d the distance between transmitter and receiver and X_σ the shadow fading term. X_σ is zero-mean Gaussian variable with standard deviation σ (in dB). The wavelength is set to the value corresponding to 60 GHz, i.e. $\lambda = 0.005$ m, and the parameters for a line-of-sight channel are set to $\gamma = 2.1$ and $\sigma = 4.9$ and for non-line-of-sight channels to $\gamma = 3.3$ and $\sigma = 7.6$.

A.4 Monotony of Achievable Rates of Leakage Based Precoding

In this appendix, we are going to show that the rate on a vertical respectively horizontal line within the region of practical interest of the transmit power - leakage power plane of leakage based precoding (see Chap. 4) is monotonously increasing for increasing leakage power respectively transmit power.

We start to show that this holds on a vertical line $\tilde{P}_{Tx}$ as shown in Fig. A.2. This can be seen by considering $\mathbf{W}_1 = \mathbf{W}_{s,1} + \mathbf{W}_{c,1}$ to be the rate optimal precoding (including the power loading on the streams) leading to a point $(\tilde{P}_{Tx}, P_{L,1})$ inside the region of practical interest. That is, no higher rate can be achieved with $\tilde{P}_{Tx}$ while generating

$P_{\mathrm{L},1}$. The leakage power can be split up into

$$P_{\mathrm{L},1} = P_{\mathrm{L,s},1} + P_{\mathrm{L,c},1} + 2P_{\mathrm{L,sc},1}, \tag{A.4}$$

with the leakage power originating from the signal term $P_{\mathrm{L,s},1} = \mathrm{tr}\left(\mathbf{W}_{\mathrm{s},1}^{\mathsf{H}}\mathbf{F}^{\mathsf{H}}\mathbf{F}\mathbf{W}_{\mathrm{s},1}\right) > 0$, the leakage power originating from the compensation term $P_{\mathrm{L,c},1} = \mathrm{tr}\left(\mathbf{W}_{\mathrm{c},1}^{\mathsf{H}}\mathbf{F}^{\mathsf{H}}\mathbf{F}\mathbf{W}_{\mathrm{c},1}\right) > 0$ and two times the leakage power of the cross term $P_{\mathrm{L,sc},1} = \mathrm{tr}\left(\mathbf{W}_{\mathrm{s},1}^{\mathsf{H}}\mathbf{F}^{\mathsf{H}}\mathbf{F}\mathbf{W}_{\mathrm{c},1}\right) = \mathrm{tr}\left(\mathbf{W}_{\mathrm{c},1}^{\mathsf{H}}\mathbf{F}^{\mathsf{H}}\mathbf{F}\mathbf{W}_{\mathrm{s},1}\right) < 0$ which is responsible for the leakage reduction. If we now construct a new precoding matrix $\mathbf{W}_2 = \alpha\mathbf{W}_{\mathrm{s},1} + \beta\mathbf{W}_{\mathrm{c},1}$, with $\alpha, \beta > 0$, such that $P_{\mathrm{Tx},2} = \tilde{P}_{\mathrm{Tx}}$ (we can always find such α and β as $P_{\mathrm{Tx},2} = \alpha^2 P_{\mathrm{s},1} + \beta^2 P_{\mathrm{c},1}$), the achievable rate is increasing as the signal term gets more power. Furthermore, the resulting leakage power

$$P_{\mathrm{L},2} = \alpha^2 P_{\mathrm{L,s},1} + \beta^2 P_{\mathrm{L,c},1} + \alpha\beta 2P_{\mathrm{L,sc},1} \tag{A.5}$$

is increased, $P_{\mathrm{L},2} > P_{\mathrm{L},1}$. This can be seen as follows. Considering only $0 < \breve{\beta} < 1$ for the moment leading to $\breve{\mathbf{W}}_2 = \mathbf{W}_{\mathrm{s},1} + \breve{\beta}\mathbf{W}_{\mathrm{c},1}$, the transmit power decreases while the rate stays constant. Thus, the leakage power has to be increased, as otherwise we could achieve equal rate with lower transmit power and lower leakage power (i.e. $\mathbf{W}_1$ would not have been rate optimal). Scaling up $\breve{\mathbf{W}}_2$ with $\alpha > 1$, $\mathbf{W}_2 = \alpha\mathbf{W}_{\mathrm{s},1} + \beta\mathbf{W}_{\mathrm{c},1}$ with $\beta = \alpha \cdot \breve{\beta}$, such that $P_{\mathrm{Tx},2} = \tilde{P}_{\mathrm{Tx}}$, we achieve a higher rate (as $\alpha > 1$) while the leakage power is increased again. Hence, for a point inside the region of practical interest with transmit power $\tilde{P}_{\mathrm{Tx}}$, we can always increase the achievable rate by allowing a higher leakage power up to the leakage power $\hat{P}_{\mathrm{L}}$ in the egoistic case.

Similarly, we can show that the achievable rate is monotonously increasing on a horizontal line in the region of practical interest. To this end, we consider $\mathbf{W}_3 = \mathbf{W}_{\mathrm{s},3} + \mathbf{W}_{\mathrm{c},3}$ to be the rate optimal precoding leading to a point $(P_{\mathrm{Tx},3}, \tilde{P}_{\mathrm{L}})$ inside the region of practical interest. We can then find a precoding matrix $\mathbf{W}_4 = \alpha\mathbf{W}_{\mathrm{s},3} + \beta\mathbf{W}_{\mathrm{c},3}$ with $\alpha, \beta > 0$, which leads to a higher achievable rate, higher transmit power $P_{\mathrm{Tx},4} > P_{\mathrm{Tx},3}$ and equal leakage power $\tilde{P}_{\mathrm{L}}$. This can be seen by considering the two following cases. If $|P_{\mathrm{L,sc},3}| > P_{\mathrm{L,c},3}$, we can find $\breve{\mathbf{W}}_4 = \mathbf{W}_{\mathrm{s},3} + \breve{\beta}\mathbf{W}_{\mathrm{c},3}$ with $\breve{\beta} > 1$, such that $\breve{P}_{\mathrm{L},4} < P_{\mathrm{L},3}$. That is, the leakage power is decreased, the transmit power is increased and the achievable rate stays constant (as $\alpha = 1$). Scaling the resulting precoding matrix with $\alpha > 1$ to $\mathbf{W}_4 = \alpha\mathbf{W}_{\mathrm{s},3} + \beta\mathbf{W}_{\mathrm{c},3}$ with $\beta = \alpha \cdot \breve{\beta}$, such that $P_{\mathrm{L},4} = \tilde{P}_{\mathrm{L}}$ then leads to an increased achievable rate (as $\alpha > 1$) and increased transmit power as desired. In the case of $|P_{\mathrm{L,sc},3}| < P_{\mathrm{L,c},3}$, we can always find $\breve{\mathbf{W}}_4 = \breve{\alpha}\mathbf{W}_{\mathrm{s},3} + \breve{\beta}\mathbf{W}_{\mathrm{c},3}$ with

$\breve{\alpha} > \breve{\beta} > 1$, such that $P_{\mathrm{L},4} < \breve{\alpha}^2 \cdot \tilde{P}_{\mathrm{L}}$. Hence, scaling the resulting precoding with $\xi < 1$, $\xi \cdot \breve{\alpha} > 1$, such that $P_{\mathrm{L},4} = \tilde{P}_{\mathrm{L}}$ then leads to an increased achievable rate (as $\xi \cdot \breve{\alpha} > 1$) and increased transmit power (as otherwise $\mathbf{W}_3$ would not have been rate optimal). The case of $|P_{\mathrm{L,sc},3}| = P_{\mathrm{L,c},3}$ was never observed in our numerical simulations and we suppose that it potentially can't occur for rate optimal precoding. Hence, for a point inside the region of practical interest with leakage power $\tilde{P}_{\mathrm{L}}$, we can always increase the achievable rate by allowing a higher transmit power up to the transmit power $\hat{P}_{\mathrm{Tx}}$ in the altruistic case.

A.5 Derivation of the Partial Derivative of the Cascade Decoding Rate

In this appendix, we are going to derive the partial derivative of R^{QF} with respect to the time τ_i allocated to relay i at the point $\tau_1 = \tau_2 = \cdots = \tau_N = 0$ given in Eq. (6.17).

The achievable decoding rate is given as

$$R^{\mathrm{QF}} = \log_2 \det\left(\mathbf{I} + (\mathbf{D}_\mathrm{z} + \mathbf{D}_\mathrm{q})^{-1}\mathbf{\Lambda}_\mathrm{s}\right). \tag{A.6}$$

If one of the $\tau_i \to 0$, the corresponding quantization noise tends to infinity

$$\sigma_{q_i}^2 = \frac{\sigma_{y_i}^2}{2^{\tau_i R_i} - 1} \to \infty, \tag{A.7}$$

and thus $1/(\sigma_{z_i}^2 + \sigma_{q_i}^2) \to 0$. Hence, the corresponding row of $\mathbf{\Lambda}_\mathrm{s}$ can be set to 0. If we now take the derivative of R^{QF} with respect to the time τ_i allocated to relay i at the point $\tau_1 = \tau_2 = \cdots = \tau_N = 0$, we can set $\tau_j = 0\ \forall j \in \{1, \ldots, N\} \setminus i$ before taking the derivative, resulting in

$$R^{\mathrm{QF}}\Big|_{\tau_j = 0\ \forall\ j \in \{1,\ldots,N\}\setminus i} = \log_2\left(1 + \frac{\lambda_{i,i}}{\sigma_{z_i}^2 + \sigma_{q_i}^2}\right). \tag{A.8}$$

Taking the derivative thereof, we get

$$\frac{\partial}{\partial \tau_i} R^{\mathrm{QF}}\Big|_{\tau_j = 0\ \forall\ j \in \{1,\ldots,N\}\setminus i} = \frac{R_i}{1 + 2^{\tau_i R_i}/\mathrm{SNR}_i}. \tag{A.9}$$

At the point $\tau_i = 0$ this simplifies to

$$\left.\frac{\partial R^{\mathrm{QF}}}{\partial \tau_i}\right|_{\tau_j=0\,\forall\, j\in\{1,\ldots,N\}} = \frac{\mathrm{SNR}_i}{1+\mathrm{SNR}_i} \cdot R_i. \tag{A.10}$$

List of Figures

2.1. Illustration of the multistage cooperative broadcast. 13

3.1. Typical setup for one hop of the cooperative broadcast. 20
3.2. Exemplary CDF of the distance from the node of interest to an arbitrary node on the disk. 22
3.3. Comparison of expected values with their approximation. 23
3.4. Comparison of coverage prediction with an actual coverage at low node density. 28
3.5. Comparison of coverage prediction with an actual coverage at high node density. 29
3.6. Empirical CDF of the distance reached in the second hop of multistage cooperative broadcast. 30
3.7. Average distance reached per hop in multistage cooperative broadcast. . 30
3.8. Average number of nodes reached per hop in the multistage cooperative broadcast: prediction compared to Monte Carlo simulations. 31
3.9. Empirical CDF of the number of nodes reached in the fourth hop of multistage cooperative broadcast. 31
3.10. Outage probability in the multistage cooperative broadcast in dependency of the coverage distance and the corresponding prediction of the coverage distance. 33
3.11. Considered setup for the multi hop transmission. 35
3.12. Average distance reached with classical multi hop transmission. 36
3.13. Empirical CDF of the distance reached in the second and fifth hop of multi hop transmission. 37
3.14. Outage probability in the multi hop transmission in dependency of the coverage distance and the corresponding prediction of the coverage distance. 38

4.1. System model for leakage based multi-user MIMO precoding. 44
4.2. The transmit power - leakage power plane. 46

4.3. Visualization of the line search in the transmit power - leakage power plane. . . . 51
4.4. Average achievable rate per mobile station. . . . 54
4.5. Average transmit and leakage power per mobile station. . . . 54
4.6. Target rate curves in the transmit power - leakage power plane. . . . 59
4.7. Illustration of the inherent coupling of the mobile stations in the network. 59
4.8. Cellular network layout for the performance evaluation of target rate precoding. . . . 61
4.9. Mutual coupling of two mobile stations for target rate precoding. . . . 63
4.10. Convergence behaviour of the achievable rates for target rate precoding. 64
4.11. CDFs of achievable rates for target rate precoding. . . . 64
4.12. Outage probabilities of target rate precoding. . . . 65
4.13. Average transmit power of target rate precoding. . . . 65

5.1. Considered scenario for beam shaping. . . . 69
5.2. Exemplary setup of a random antenna array. . . . 70
5.3. Exemplary setup for leakage based beam shaping. . . . 72
5.4. Exemplary radiation patterns of leakage based beam shaping. . . . 75
5.5. Transmit power - leakage power trade-off for leakage based beam shaping. 76
5.6. Impact of the suppression factor c. . . . 77
5.7. Achievable gains of leakage based beam shaping. . . . 79
5.8. Average main-to-sidelobe-gain ratio of leakage based beam shaping. . . 81
5.9. Average sorted transmit power per node of leakage based beam shaping. 82
5.10. Impact of erroneous location information on leakage based beam shaping. 84

6.1. System setup of quantize-and-forward virtual MIMO receive cooperation. 91
6.2. Average decoding rate R^{QF} in regime II. . . . 98
6.3. Total average throughput R_{MS} in regime II. . . . 99
6.4. Total average throughput R_{MS} in regime III. . . . 100
6.5. Instantaneous τ_i at the respective peak performance for one specific channel realization in regime III. . . . 101
6.6. Instantaneous total throughput at the instantaneous optimal τ_{tot} and N_{o}, and at the average optimal τ_{tot} and N_{o} in regime III. . . . 101
6.7. Number of operating relays and total second hop duration for the cascade resource allocation algorithm and the gradient search in regime III. . . 105

6.8. Achievable decoding rate R^{QF} and achievable throughput R_{MS} for the cascade resource allocation algorithm and the gradient search in regime III. 106

7.1. Setup for the performance comparison of multi hop transmission, multistage cooperative broadcast and transmit cooperation with beam shaping. 111
7.2. Reachable distance for beam shaping, multistage cooperative broadcast and multi hop transmission. 114
7.3. Number of required time slots to serve all destinations. 118
7.4. Outage probability at destinations for different source-destination distances. 118
7.5. Final throughput of the virtual MIMO scheme compared to sequentially serving the destinations. 119

8.1. Urban hotspot scenario. 123
8.2. Illustration of the virtual MIMO transmit cooperation. 129
8.3. Uplink: Performance in the access phase. 130
8.4. Uplink: Performance in the exchange phase. 131
8.5. Uplink: Average achievable mobile station sum rate. 132
8.6. Uplink: Achievable mobile station sum rate for variable bandwidth scaling factors. 133
8.7. Illustration of the quantize-and-forward virtual MIMO receive cooperation. 134
8.8. Downlink: Number of operating mobile stations which forward their observation, and total duration of exchange phase. 136
8.9. Downlink: Average achievable decoding rate. 137
8.10. Downlink: Average system throughput. 137
8.11. Downlink: Average system throughput for varying bandwidth scaling factor. 138
8.12. Illustration of one round of the reuse protocol. 140
8.13. Reuse patterns. 141
8.14. Uplink: Performance in access phase with spatial reuse. 142
8.15. Uplink: Performance in exchange phase with spatial reuse. 143
8.16. Uplink: Average system throughput with spatial reuse. 143
8.17. Uplink: Performance of local users. 144
8.18. Uplink: Average system throughput with spatial reuse for varying bandwidth scaling factor. 145

A.1. Antenna pattern as considered in the cellular network setup. 150
A.2. The transmit power - leakage power plane. 151

Acronyms

bpcu	Bits per channel use.
bpfhcu	Bits per first hop channel use.
BPP	Binomial point process.
BS	Base station.
CDF	Cumulative distribution function.
CoMP	Coordinated multipoint.
CRAN	Cloud radio access network.
CSMA/CA	Carrier sense multiple access/collision avoidance.
DSDV	Destination-sequenced distance-vector.
DSR	Dynamic source routing.
GEVD	Generalized eigenvalue decomposition.
GPS	Global positioning system.
i.i.d.	Independent and identically distributed.
ISM	Industrial, scientific and medical.
LU	Local user.
MANET	Mobile ad hoc network.
MIMO	Multiple-input multiple-output.
mmW	Millimeter wave.
MOI	Mobile of interest.
MS	Mobile station.

PDF	Probability density function.
PPP	Poisson point process.
RBAP	Residential backhaul access point.
SLNR	Signal-to-leakage-plus-noise ratio.
SNR	Signal-to-noise ratio.
SQNR	Signal-to-quantization-noise ratio.
TxMF	Transmit matched filter.
UHF	Ultra high frequency.
ULA	Uniform linear array.
URA	Uniform rectangular array.
UWB	Ultra wide band.
VHF	Very high frequency.
VM	Victim mobile.
WLAN	Wireless local area network.
ZRP	Zone routing protocol.

Notation and Variables

Notation

a, A	Scalars a and A.
$\mathbf{a}$	Vector $\mathbf{a}$.
$\mathbf{A}$	Matrix $\mathbf{A}$.
$\det(\mathbf{A})$	Determinant of matrix $\mathbf{A}$.
$\mathrm{tr}(\mathbf{A})$	Trace of matrix $\mathbf{A}$.
$\mathbf{A}^{\mathsf{T}}$	Transpose of matrix $\mathbf{A}$.
$\mathbf{A}^{\mathsf{H}}$	Hermitian transpose of matrix $\mathbf{A}$.
$\mathrm{GEVD}(\mathbf{A}, \mathbf{B})$	Generalized eigenvalue decomposition of matrices $\mathbf{A}$ and $\mathbf{B}$.
$\mathsf{E}[X]$	Expected value of random variable X.
$\lceil a \rceil$	a rounded to the next bigger integer value.
$\mathcal{CN}(\mu, \sigma^2)$	Circularly symmetric complex Gaussian distribution with mean μ and variance σ^2.
$\mathcal{U}(a, b)$	Uniform distribution on interval [a,b].
$\mathbf{I}$	Identity matrix.

Selected variables

Chapter 2:

η	Average number of transmissions.
N	Number of nodes.
ξ	Ratio of coverage area and total network area.

Chapter 3:

δ	Node density.

d	Distance.
d_n	Distance to n-th closest node.
$d_{\max,k}$	Coverage range in k-th hop.
D	Distance from center of disk.
γ	Path loss coefficient.
h	Fading variable.
N	Number of nodes.
N_k	Number of covered nodes in hop k.
P_{Tx}	Transmit power.
ρ	Radius of disk around source.
r	Radius of circle around node of interest.
$R_{\min}$	Minimal required achievable rate.
σ_w^2	Noise variance of additive white Gaussian noise.
U_n	CDF value of d_n.

Chapter 4:

$\mathbf{D}$	Diagonal matrix with generalized eigenvalues.
η	Tolerance value.
$\mathbf{F}$	Channel to victim mobiles.
$\mathbf{H}$	Channel to mobile of interest.
$\mathbf{K}_{\mathrm{z}}$	Interference plus noise covariance matrix.
$\mathbf{\Lambda}_{\mathrm{s}}$	Transmit signal covariance matrix.
N_{BS}	Number of base station antennas.
N_{MS}	Number of mobile station antennas.
N_{VM}	Number of victim mobile antennas.
P_{c}	Transmit power allocated to compensation term.
P_{L}	Leakage power.
$\tilde{P}_{\mathrm{L}}$	Leakage power constraint.
P_{s}	Transmit power allocated to desired signal.
P_{Tx}	Transmit power.
$\tilde{P}_{\mathrm{Tx}}$	Transmit power constraint.
P_{ϑ}	Weighted sum power.

$\tilde{P}_\vartheta$	Weighted sum power constraint.
$\mathbf{Q}$	Transmit covariance matrix.
R_t	Target rate.
σ_w^2	Noise variance of additive white Gaussian noise.
ϑ	Weighting variable.
$\mathbf{W}$	Precoding matrix.
$\mathbf{W}_\mathrm{c}$	Compensation term of precoding matrix.
$\mathbf{W}_\mathrm{s}$	Signal term of precoding matrix.
$\mathbf{Z}$	Matrix with generalized eigenvectors.

Chapter 5:

a	Sidelength of transmit cluster.
c	Suppression factor.
$\mathbf{C}_\mathrm{F}$	Weighting matrix for undesired directions.
$\mathbf{C}_\mathrm{H}$	Weighting matrix for desired directions.
d_i	Path length difference of node i.
$\mathbf{F}$	Channel in undesired directions.
$\mathbf{H}$	Channel in desired directions.
λ	Wavelength.
M	Number of look directions.
N	Number of nodes.
ϕ	Look direction.
ϕ_s	Desired direction.
ϕ_L	Direction of sector L.
ψ_s	Beam width centered around desired direction.
ψ_l	Width of sector L.
P_L	Leakage power.
P_s	Signal power in desired directions.
P_ϑ	Weighted sum power.
P_Tx	Transmit power.
ϑ	Weighting variable.
$\bar{\mathbf{v}}$	Steering vector.

$\mathbf{V}$	Matrix with steering vectors.
$\mathbf{w}$	Weight vector.

Chapter 6:

$\mathbf{D}_\mathrm{q}$	Quantization noise covariance matrix.
$\mathbf{D}_\mathrm{z}$	Interference plus noise covariance matrix.
$h_{i,j}$	Channel from source j to relay i.
$\mathbf{H}$	Channel between sources and relays.
$\lambda_{i,i}$	Received signal power at relay i.
$\mathbf{\Lambda}_\mathrm{s}$	Received signal covariance matrix.
N	Number of sources/relays.
N_o	Number of operating relays.
P_Tx	Transmit power of relays.
R_i	Backhaul rate of relay i.
Q_i	Effective quantization rate of relay i.
R_MS	Overall throughput.
R^QF	Achievable decoding rate.
$\sigma_{q_i}^2$	Quantization noise variance of relay i.
σ_w^2	Noise variance.
$\sigma_{y_i}^2$	Variance of received signal at relay i.
$\sigma_{z_i}^2$	Interference plus noise variance at relay i.
$\mathbf{snr}$	SNR vector.
SNR_i	Signal-to-noise ratio at relay i.
SQNR_i	Signal-to-quantization-noise ratio at relay i.
τ_i	Channel access time of relay i.
τ_tot	Total second hop duration.

Chapter 7:

δ	Node density.
d	Distance.
d_0	Reference distance.
$d_{\mathrm{max},k}$	Coverage range in hop k.

D	Distance from source.
γ	Path loss coefficient.
h	Channel coefficient.
K	Unitless constant.
N	Number of nodes.
P_{s}	Received signal power.
P_{Tx}	Transmit power.
R_{min}	Transmission rate.
P_{out}	Outage probability.
R_{MS}	Throughput.
σ_w^2	Noise variance.
$\mathrm{SNR}_{\mathrm{min}}$	Minimal required SNR.
R_{EX}	Exchange rate.
ξ	Scaling factor for R_{EX}.

Chapter 8:

a	Sidelength of setup.
b	Sidelength of urban hotspot.
β	Bandwidth scaling factor.
$\lambda_{i,i}$	Received signal power at mobile station i.
$\mathbf{\Lambda}_{\mathrm{s}}$	Received signal covariance matrix.
M	Number of nodes in sub-cluster.
N_{AP}	Number of active RBAPs.
$\bar{N}_{\mathrm{AP}}$	Number of inactive RBAPs.
N_{LU}	Number of local users.
N_{MS}	Number of active mobile stations.
$\bar{N}_{\mathrm{MS}}$	Number of inactive mobile stations.
P_{EX}	Transmit power in local exchange.
P_{LU}	Transmit power of local users.
P_{MS}	Transmit power of mobile stations to RBAPs.
Q_i	Effective quantization rate of mobile station i.
r^{AC}	Reuse factor in access phase.

r^{EX}	Reuse factor in exchange phase.
R_{b}	Averaged backhaul rate.
R_i	Broadcasting rate of mobile station i.
R_{LU}	Averaged sum of local users.
R_{MS}	Overall throughput.
$R_{\mathrm{MS},c}^{\mathrm{AC}}$	Achievable rate of sub-cluster c in access phase.
$R_{\mathrm{MS,tot}}^{\mathrm{AC}}$	Sum of achievable rates in access phase over all sub-clusters.
R^{QF}	Achievable decoding rate.
σ_w^2	Noise variance.
$\sigma_{q_i}^2$	Quantization noise variance of mobile station i.
$\sigma_{z_i}^2$	Interference plus noise variance at mobile station i.
τ_i	Channel access time of mobile station i.
$\tau_{\max,i}^{\mathrm{EX}}$	Time to share data in all sub-cluster of the current set.
τ_{tot}	Total second hop duration.
$\tau_{\mathrm{tot},c}^{\mathrm{EX}}$	Time to share data in sub-cluster c.
$\mathbf{W}$	Precoding matrix.

Bibliography

[1] Peter Jonsson and others, "Ericsson Mobility Report," (2017). [Online]. Available: https://www.ericsson.com/en/mobility-report/reports/november-2017

[2] S. Al-Sultan, M. M. Al-Doori, A. H. Al-Bayatti, and H. Zedan, "A comprehensive survey on vehicular ad hoc network," *Journal of Network and Computer Applications*, vol. 37, pp. 380 – 392, 2014. [Online]. Available: http://www.sciencedirect.com/science/article/pii/S108480451300074X

[3] J. Lin, W. Yu, N. Zhang, X. Yang, H. Zhang, and W. Zhao, "A survey on internet of things: Architecture, enabling technologies, security and privacy, and applications," *IEEE Internet of Things Journal*, vol. 4, no. 5, pp. 1125–1142, Oct 2017.

[4] G. J. Foschini, "Layered space-time architecture for wireless communication in a fading environment when using multi-element antennas," *Bell Labs Technical Journal*, vol. 1, no. 2, pp. 41–59, Autumn 1996.

[5] E. Telatar, "Capacity of multi-antenna gaussian channels," *European Trans. on Telecommunications*, vol. 10, no. 6.

[6] S. M. Alamouti, "A simple transmit diversity technique for wireless communications," *IEEE Journal on Selected Areas in Communications*, vol. 16, no. 8, pp. 1451–1458, Oct 1998.

[7] V. Tarokh, N. Seshadri, and A. R. Calderbank, "Space-time codes for high data rate wireless communication: performance criterion and code construction," *IEEE Transactions on Information Theory*, vol. 44, no. 2, pp. 744–765, Mar 1998.

[8] J. Mietzner, R. Schober, L. Lampe, W. H. Gerstacker, and P. A. Hoeher, "Multiple-antenna techniques for wireless communications - a comprehensive literature survey," *IEEE Communications Surveys Tutorials*, vol. 11, no. 2, pp. 87–105, Second 2009.

[9] R. J. Mailloux, *Phased Array Antenna Handbook*. Artech House, 2005.

[10] L. C. Godara, "Applications of antenna arrays to mobile communications. i. performance improvement, feasibility, and system considerations," *Proceedings of the IEEE*, vol. 85, no. 7, pp. 1031–1060, Jul 1997.

[11] ——, "Application of antenna arrays to mobile communications. ii. beam-forming and direction-of-arrival considerations," *Proceedings of the IEEE*, vol. 85, no. 8, pp. 1195–1245, Aug 1997.

[12] D. Gesbert, M. Shafi, D. shan Shiu, P. J. Smith, and A. Naguib, "From theory to practice: an overview of MIMO space-time coded wireless systems," *IEEE Journal on Selected Areas in Communications*, vol. 21, no. 3, pp. 281–302, Apr 2003.

[13] D. G. Brennan, "Linear diversity combining techniques," *Proceedings of the IEEE*, vol. 91, no. 2, pp. 331–356, Feb 2003.

[14] G. Foschini and M. Gans, "On limits of wireless communications in a fading environment when using multiple antennas," *Wireless Personal Communications*, vol. 6, no. 3, pp. 311–335, Mar 1998.

[15] Q. H. Spencer, C. B. Peel, A. L. Swindlehurst, and M. Haardt, "An introduction to the multi-user mimo downlink," *IEEE Communications Magazine*, vol. 42, no. 10, pp. 60–67, Oct 2004.

[16] S. Schwarz and M. Rupp, "Exploring coordinated multipoint beamforming strategies for 5g cellular," *IEEE Access*, vol. 2, pp. 930–946, 2014.

[17] Q. Spencer, A. Swindlehurst, and M. Haardt, "Zero-forcing methods for downlink spatial multiplexing in multiuser MIMO channels," *Signal Processing, IEEE Transactions on*, vol. 52, no. 2, pp. 461–471, 2004.

[18] M. Sadek, A. Tarighat, and A. Sayed, "A leakage-based precoding scheme for downlink multi-user MIMO channels," *IEEE Trans. on Wireless Communications*, vol. 6, no. 5, pp. 1711–1721, 2007.

[19] L. Zheng and D. N. C. Tse, "Diversity and multiplexing: a fundamental tradeoff in multiple-antenna channels," *IEEE Transactions on Information Theory*, vol. 49, no. 5, pp. 1073–1096, May 2003.

[20] A. Nosratinia, T. E. Hunter, and A. Hedayat, "Cooperative communication in wireless networks," *IEEE Communications Magazine*, vol. 42, no. 10, pp. 74–80, Oct 2004.

[21] B. Sirkeci-Mergen, A. Scaglione, and G. Mergen, "Asymptotic analysis of multistage cooperative broadcast in wireless networks," *IEEE Transactions on Information Theory*, vol. 52, no. 6, pp. 2531–2550, June 2006.

[22] M. Dohler, "Virtual antenna arrays," *Ph.D. dissertation, King's College London, Univ. London*, Nov 2003.

[23] A. Ozgur, O. Leveque, and D. Tse, "Hierarchical cooperation achieves optimal capacity scaling in ad hoc networks," *Information Theory, IEEE Transactions on*, vol. 53, no. 10, pp. 3549–3572, Oct 2007.

[24] H. Zhao, Y. Xi, J. Wei, and J. Li, "A typical cooperative MIMO scheme in wireless ad hoc networks and its channel capacity," in *Communications, 2008. ICC '08. IEEE Int. Conference on*, May 2008, pp. 4644–4648.

[25] J. Loo, J. Mauri, and J. Ortiz, *Mobile Ad Hoc Networks.* Auerbach Publications, 2012.

[26] X. Li, *Wireless Ad Hoc and Sensor Networks.* Cambridge University Press, 2008.

[27] C. Zhang, X. Zhu, and Y. Fang, "On the improvement of scaling laws for large-scale MANETs with network coding," *IEEE Journal on Selected Areas in Communications*, vol. 27, no. 5, June 2009.

[28] H. Xu, X. Wu, H. R. Sadjadpour, and J. J. Garcia-Luna-Aceves, "A unified analysis of routing protocols in MANETs," *IEEE Transactions on Communications*, vol. 58, no. 3, March 2010.

[29] F. Gentile, T. Rüegg, M. Kuhn, and A. Wittneben, *Routing Protocols for MANETs: Overhead Evaluation in Military MANETs.* Technical report, available upon request, 2017.

[30] F. Rusek *et al.*, "Scaling up MIMO: Opportunities and challenges with very large arrays," *Signal Processing Magazine, IEEE*, vol. 30, no. 1, pp. 40–60, Jan 2013.

[31] A. Ghosh, T. A. Thomas, M. C. Cudak, R. Ratasuk, P. Moorut, F. W. Vook, T. S. Rappaport, G. R. MacCartney, S. Sun, and S. Nie, "Millimeter-wave enhanced local area systems: A high-data-rate approach for future wireless networks," *IEEE Journal on Selected Areas in Communications*, vol. 32, no. 6, pp. 1152–1163, June 2014.

[32] I. Hwang, B. Song, and S. Soliman, "A holistic view on hyper-dense heterogeneous and small cell networks," *Communications Magazine, IEEE*, vol. 51, no. 6, pp. 20–27, June 2013.

[33] T. Rüegg, F. Gentile, and A. Wittneben, "Cooperative broadcast performance prediction based on inter-node distance distributions," in *IEEE Wireless Communications and Networking Conference (WCNC) 2018*, Apr. 2018.

[34] T. Rüegg, R. Rolny, and A. Wittneben, "Leakage based beam shaping for cooperative communication in MANETs," in *International ITG Workshop on Smart Antennas*, Mar. 2018.

[35] T. Rüegg, Y. Hassan, and A. Wittneben, "Low complexity resource allocation for QF VMIMO receivers with a shared backhaul," in *IEEE International Symposium on Personal, Indoor and Mobile Radio Communications (PIMRC)*, Oct. 2017.

[36] ——, "User cooperation enabled traffic offloading in urban hotspots," in *IEEE International Symposium on Personal, Indoor and Mobile Radio Communications (PIMRC)*, Sep. 2016.

[37] T. Rüegg and A. Wittneben, "User cooperation for traffic offloading in remote hotspots," in *International ITG Workshop on Smart Antennas*, Mar. 2016.

[38] ——, "Resource allocation for QF VMIMO receive cooperation in urban traffic hotspots," in *26th European Signal Processing Conference (EUSIPCO)*, Sep. 2018.

[39] T. Rüegg, A. U. T. Amah, and A. Wittneben, "On the trade-off between transmit and leakage power for rate optimal MIMO precoding," in *IEEE Workshop on Signal Processing Advances for Wireless Communications (SPAWC)*, Jun. 2014.

[40] T. Rüegg, M. Kuhn, and A. Wittneben, "Decentralized target rate optimization for MU-MIMO leakage based precoding," in *Asilomar Conference on Signals, Systems, and Computers*, Nov. 2014.

[41] M. Abolhasan, T. Wysocki, and E. Dutkiewicz, "A review of routing protocols for mobile ad hoc networks," *Ad Hoc Networks*, vol. 2, no. 1, pp. 1 – 22, 2004. [Online]. Available: http://www.sciencedirect.com/science/article/pii/S157087050300043X

[42] B. Williams and T. Camp, "Comparison of broadcasting techniques for mobile ad hoc networks," in *Proceedings of the 3rd ACM International Symposium on Mobile Ad Hoc Networking &Amp; Computing*, ser. MobiHoc '02. New York, NY, USA: ACM, 2002, pp. 194–205. [Online]. Available: http://doi.acm.org/10.1145/513800.513825

[43] P. Mitran, H. Ochiai, and V. Tarokh, "Space-time diversity enhancements using collaborative communications," *Information Theory, IEEE Transactions on*, vol. 51, no. 6, pp. 2041–2057, June 2005.

[44] I. Hammerstrom, M. Kuhn, and A. Wittneben, "Cooperative diversity by relay phase rotations in block fading environments," in *IEEE SPAWC, 2004.*, July 2004, pp. 293–297.

[45] Q. M. Tran and A. Dadej, "Proactive routing overhead in mobile ad-hoc networks," in *Proceeding of IEEE International Symposium on a World of Wireless, Mobile and Multimedia Networks 2014*, June 2014, pp. 1–6.

[46] K. Vardhe and D. Reynolds, "On the performance of multistage cooperative networks," in *2010 48th Annual Allerton Conference on Communication, Control, and Computing (Allerton)*, Sept 2010, pp. 909–914.

[47] Z. Mobini and M. Khabbazian, "Asymptotic gain analysis of cooperative broadcast in linear wireless networks," *IEEE Transactions on Wireless Communications*, vol. 15, no. 1, pp. 485–497, Jan 2016.

[48] Z. Ding, K. K. Leung, D. L. Goeckel, and D. Towsley, "Cooperative transmission protocols for wireless broadcast channels," *IEEE Transactions on Wireless Communications*, vol. 9, no. 12, pp. 3701–3713, December 2010.

[49] M. Haenggi, J. G. Andrews, F. Baccelli, O. Dousse, and M. Franceschetti, "Stochastic geometry and random graphs for the analysis and design of wireless networks," *IEEE Journal on Selected Areas in Communications*, vol. 27, no. 7, pp. 1029–1046, September 2009.

[50] S. Srinivasa and M. Haenggi, "Distance distributions in finite uniformly random networks: Theory and applications," *IEEE Transactions on Vehicular Technology*, vol. 59, no. 2, pp. 940–949, Feb 2010.

[51] V. Naghshin, M. C. Reed, and Y. Liu, "On the performance analysis of finite wireless network," in *2015 IEEE International Conference on Communications (ICC)*, June 2015, pp. 6530–6535.

[52] A. Thornburg, T. Bai, and R. W. Heath, "Performance analysis of outdoor mmwave ad hoc networks," *IEEE Transactions on Signal Processing*, vol. 64, no. 15, pp. 4065–4079, Aug 2016.

[53] H. A. David and H. N. Nagaraja, *Order Statistics*, 3rd ed. John Wiley & Sons, 2003.

[54] B. O. Lee, H. W. Je, I. Sohn, O. Shin, and K. B. Lee, "Interference-aware decentralized precoding for multicell MIMO TDD systems," in *IEEE GLOBECOM 2008 - 2008 IEEE Global Telecommunications Conference*, Nov 2008, pp. 1–5.

[55] Patcharamaneepakorn *et al.*, "Weighted sum capacity maximization using a modified leakage-based transmit filter design," *IEEE Trans. on Vehicular Technology*, vol. 62, no. 3, pp. 1177–1188, March 2013.

[56] P. Cheng, M. Tao, and W. Zhang, "A new SLNR-based linear precoding for downlink multi-user multi-stream MIMO systems," *Communications Letters, IEEE*, vol. 14, no. 11, pp. 1008–1010, November 2010.

[57] F. Xinxin, M. Wang, W. Yaxi, F. Haiqiang, and L. Jinhui, "An efficient power allocation scheme for leakage-based precoding in multi-cell multiuser MIMO downlink," *Communications Letters, IEEE*, vol. 15, no. 10, pp. 1053–1055, October 2011.

[58] M. Schubert and H. Boche, "Solution of the multiuser downlink beamforming problem with individual sinr constraints," *IEEE Transactions on Vehicular Technology*, vol. 53, no. 1, pp. 18–28, Jan 2004.

[59] S. Han, C. Yang, G. Wang, D. Zhu, and M. Lei, "Coordinated multi-point transmission strategies for TDD systems with non-ideal channel reciprocity," *IEEE Transactions on Communications*, vol. 61, no. 10, pp. 4256–4270, October 2013.

[60] E. Bjornson, R. Zakhour, D. Gesbert, and B. Ottersten, "Cooperative multicell precoding: Rate region characterization and distributed strategies with instantaneous and statistical CSI," *Signal Processing, IEEE Transactions on*, vol. 58, no. 8, pp. 4298–4310, 2010.

[61] Y. Wu, J. Zhang, M. Xu, S. Zhou, and X. Xu, "Multiuser mimo downlink precoder design based on the maximal SJNR criterion," in *Global Telecommunications Conference, 2005. GLOBECOM '05. IEEE*, vol. 5, Dec 2005, pp. 5 pp.–2698.

[62] J. Park, G. Lee, Y. Sung, and M. Yukawa, "Coordinated beamforming with relaxed zero forcing: The sequential orthogonal projection combining method and rate control," *IEEE Transactions on Signal Processing*, vol. 61, no. 12, pp. 3100–3112, June 2013.

[63] C.-H. Fung, W. Yu, and T. J. Lim, "Precoding for the multiantenna downlink: Multiuser SNR gap and optimal user ordering," *Communications, IEEE Transactions on*, vol. 55, no. 1, pp. 188–197, Jan 2007.

[64] M. Mohseni, R. Zhang, and J. Cioffi, "Optimized transmission for fading multiple-access and broadcast channels with multiple antennas," *Sel. Areas in Comm., IEEE Journal on*, vol. 24, no. 8, pp. 1627–1639, Aug 2006.

[65] C. Hellings, M. Joham, and W. Utschick, "Gradient-based power minimization in MIMO broadcast channels with linear precoding," *Signal Processing, IEEE Transactions on*, vol. 60, no. 2, pp. 877–890, Feb 2012.

[66] J. Löfberg, "Yalmip : A toolbox for modeling and optimization in MATLAB," in *Proceedings of the CACSD Conference*, 2004.

[67] M. Kuhn, R. Rolny, A. Wittneben, M. Kuhn, and T. Zasowski, "The potential of restricted PHY cooperation for the downlink of LTE-advanced," in *Vehicular Technology Conference, 2011 IEEE*, Sept 2011, pp. 1–5.

[68] 3rd Generation Partnership Project 3GPP, "Further advancements for E-UTRA physical layer aspects (release 9)," *3GPP TR 36.814, V9.0.0*, March 2010.

[69] P. Kyösti *et al.*, "WINNER II channel models," *Tech. Rep. IST-4-027759 WINNER II D1.1.2 V1.2 Part I Channel Models*, Sep 2007.

[70] M. Sadek, A. Tarighat, and A. H. Sayed, "Active antenna selection in multiuser mimo communications," *IEEE Transactions on Signal Processing*, vol. 55, no. 4, pp. 1498–1510, April 2007.

[71] C. A. Balanis, *Antenna Theory.* Wiley-Interscience, 2005.

[72] S. P. Applebaum, "Adaptive arrays," *IEEE Transactions on Antennas and Propagation*, vol. 24, no. 5, pp. 585–598, September 1976.

[73] C. A. Olen and J. R. T. Compton, "A numerical pattern synthesis algorithm for arrays," *IEEE Transactions on Antennas and Propagation*, vol. 38, no. 10, pp. 1666–1676, October 1990.

[74] P. Y. Zhou and M. A. Ingram, "Pattern synthesis for arbitrary arrays using an adaptive array method," *IEEE Transactions on Antennas and Propagation*, vol. 47, no. 5, pp. 862–869, May 1999.

[75] R. L. Haupt, "Adaptive antenna arrays using a genetic algorithm," in *2006 IEEE Mountain Workshop on Adaptive and Learning Systems*, July 2006, pp. 249–254.

[76] S. Zhao, Y. Chen, and J. A. Farrell, "High-precision vehicle navigation in urban environments using an MEM's IMU and single-frequency GPS receiver," *IEEE Transactions on Intelligent Transportation Systems*, vol. 17, no. 10, pp. 2854–2867, Oct 2016.

[77] Z. W. Mekonnen and A. Wittneben, "Self-calibration method for TOA based localization systems with generic synchronization requirement," in *IEEE ICC 2015, London, UK*, Jun. 2015.

[78] A. J. Weiss and B. Friedlander, "Array shape calibration using sources in unknown locations-a maximum likelihood approach," *IEEE Transactions on Acoustics, Speech, and Signal Processing*, vol. 37, no. 12, pp. 1958–1966, Dec 1989.

[79] Y. Hassan, B. Gahr, and A. Wittneben, "Rate maximization in dense interference networks using non-cooperative passively loaded relays," in *Asilomar Conference on Signals, Systems, and Computers*, Nov. 2015. [Online]. Available: http://www.nari.ee.ethz.ch/wireless//pubs/p/AsilomarYahia15

[80] A. Sanderovich, S. Shamai, and Y. Steinberg, "Distributed MIMO receiver - achievable rates and upper bounds," *IEEE Transactions on Information Theory*, vol. 55, no. 10, pp. 4419–4438, Oct 2009.

[81] R. Karasik, O. Simeone, and S. S. Shitz, "Robust uplink communications over fading channels with variable backhaul connectivity," *IEEE Transactions on Wireless Communications*, vol. 12, no. 11, pp. 5788–5799, November 2013.

[82] L. Zhou and W. Yu, "Uplink multicell processing with limited backhaul via per-base-station successive interference cancellation," *IEEE Journal on Sel. Areas in Com.*, vol. 31, no. 10, pp. 1981–1993, October 2013.

[83] Y. Zhou and W. Yu, "Optimized backhaul compression for uplink cloud radio access network," *IEEE Journal on Selected Areas in Communications*, vol. 32, no. 6, pp. 1295–1307, June 2014.

[84] Y. Zhou, Y. Xu, W. Yu, and J. Chen, "On the optimal fronthaul compression and decoding strategies for uplink cloud radio access networks," *IEEE Trans. on Inf. Theory*, vol. 62, no. 12, pp. 7402–7418, Dec 2016.

[85] T. M. Cover and J. A. Thomas, *Elements of Information theory*, 2nd ed. Wiley Interscience, 2006.

[86] A. Goldsmith, *Wireless Communications.* Cambridge University Press, 2005.

[87] armasuisse Science and Technology, *Personal communications.* Swiss confederation, Federal Department of Defence, Civil Protection and Sport DDPS, March 2017.

[88] S. Rangan, T. Rappaport, and E. Erkip, "Millimeter-wave cellular wireless networks: Potentials and challenges," *Proceedings of the IEEE*, vol. 102, no. 3, pp. 366–385, March 2014.

[89] O. Simeone, O. Somekh, G. Kramer, H. Poor, and S. Shamai, "Throughput of cellular systems with conferencing mobiles and cooperative base stations," *"EURASIP" Journal on Wireless Communications and Networking*, vol. 2008, no. 10, pp. 3549–3572, Feb 2008.

[90] R. Rolny, "Future mobile communication: From cooperative cells to the post-cellular relay carpet," Ph.D. dissertation, ETH Zürich, 2016.

[91] 3rd Generation Partnership Project 3GPP, "Spatial channel model for multiple input multiple output (MIMO) simulations," *3GPP TR 25.996, V12.0.0*, Sep 2014.

[92] T. A. Thomas and F. W. Vook, "System level modeling and performance of an outdoor mmwave local area access system," in *2014 IEEE 25th Annual International Symposium on Personal, Indoor, and Mobile Radio Communication (PIMRC)*, Sept 2014, pp. 108–112.

Curriculum Vitae

Name:	**Tim Rüegg**
Date of Birth:	March 16, 1987
Citizen:	Eschenbach SG, Switzerland
Nationality:	Swiss

Education

08/2012-08/2018	**ETH Zurich, Switzerland** PhD studies at the Communication Technology Laboratory, Department of Information Technology and Electrical Engineering.
09/2016-06/2018	**ETH Zurich, Switzerland** Teaching Certificate in Information Technology and Electrical Engineering.
09/2006-04/2012	**ETH Zurich, Switzerland** Studies in Information Technology and Electrical Engineering. Degree: Master of Science (MSc. ETH), with distinction.
08/2009-12/2009	**University of California Berkely, USA** Visiting scholar in the Berkeley-ETHZ exchange program.
08/2001-06/2005	**Kantonsschule Schaffhausen, Switzerland** Matura with focus on applied mathematics and physics.

Experience

08/2012-10/2018	**ETH Zurich, Switzerland** Research assistant at the Communication Technology Laboratory headed by Prof. Dr. Armin Wittneben. • Research in wireless communications; • Project work in an industry project with armasuisse Science and Technology; • Teaching and presentation experience; • Supervision of master's thesis and student projects; • Organization of the annual "International Seminar on Mobile Communications"; • Review of international conference submissions.
09/2011-12/2011	**ETH Zurich, Switzerland** Teaching Assistant for the lecture "Signals and Systems I".
02/2010-06/2010	**Rheinmetall Air Defence AG, Zurich** Industrial Internship: Improvement of RADAR search patterns and track-while-scan solutions.

Publications

- **Journal Papers**
  - **The Cellular Relay Carpet: Distributed Cooperation with Ubiquitous Relaying**
    R. Rolny, T. Rüegg, M. Kuhn, and A. Wittneben, Springer International Journal of Wireless Information Networks, June 2014.
- **Conference, Symposium, and Workshop Papers**
  - **Resource Allocation for QF VMIMO Receive Cooperation in Urban Traffic Hotspots**
    T. Rüegg and A. Wittneben, 26th European Signal Processing Conference (EUSIPCO), Rome, Italy, September 2018.
  - **Cooperative Broadcast Performance Prediction Based on Inter-Node Distance Distributions**
    T. Rüegg, F. Gentile, and A. Wittneben, IEEE Wireless Communications and Networking Conference (WCNC), Barcelona, Spain, April 2018.
  - **Leakage Based Beam Shaping for Cooperative Communication in MANETs**
    T. Rüegg, R. Rolny, and A. Wittneben, International ITG Workshop on Smart Antennas, Bochum, Germany, March 2018.
  - **Low Complexity Resource Allocation for QF VMIMO Receivers with a Shared Backhaul**

T. Rüegg, Y. Hassan, and A. Wittneben, IIEEE International Symposium on Personal, Indoor and Mobile Radio Communications (PIMRC), Montreal, Canada, October 2017.

- **User Cooperation Enabled Traffic Offloading in Urban Hotspots**
  T. Rüegg, Y. Hassan, and A. Wittneben, IIEEE International Symposium on Personal, Indoor and Mobile Radio Communications (PIMRC), Valencia, Spain, September 2016.
- **User Cooperation for Traffic Offloading in Remote Hotspots**
  T. Rüegg, and A. Wittneben, International ITG Workshop on Smart Antennas, Munich, Germany, March 2016.
- **Decentralized Target Rate Optimization for MU-MIMO Leakage Based Precoding**
  T. Rüegg, M. Kuhn, and A. Wittneben, Asilomar Conference on Signals, Systems, and Computers, Pacific Grove, USA, November 2014.
- **On the Trade-Off Between Transmit and Leakage Power for Rate Optimal MIMO Precoding**
  T. Rüegg, A. U. T. Amah, and A. Wittneben, IEEE Workshop on Signal Processing Advances for Wireless Communications (SPAWC), Toronto, Canada, June 2014.

Bisher erschienene Bände der Reihe

Series in Wireless Communications

ISSN 1611-2970

1	Dirk Benyoucef	Codierte Multiträgerverfahren für Powerline- und Mobilfunkanwendungen ISBN 978-3-8325-0137-2 40.50 EUR
2	Marc Kuhn	Space-Time Codes und ihre Anwendungen in zukünftigen Kommunikationssystemen ISBN 978-3-8325-0159-4 40.50 EUR
3	Frank Althaus	A New Low-Cost Approach to Wireless Communication over Severe Multipath Fading Channels ISBN 978-3-8325-0158-7 40.50 EUR
4	Ingmar Hammerström	Cooperative Relaying and Adaptive Scheduling for Low Mobility Wireless Access Networks ISBN 978-3-8325-1466-2 40.50 EUR
5	Thomas Zasowski	A System Concept for Ultra Wideband (UWB) Body Area Networks ISBN 978-3-8325-1715-1 40.50 EUR
6	Boris Rankov	Spectral Efficient Cooperative Relaying Strategies for Wireless Networks ISBN 978-3-8325-1744-1 40.50 EUR
7	Stephan Sand	Joint Iterative Channel and Data Estimation in High Mobility MIMO-OFDM Systems ISBN 978-3-8325-2385-5 39.00 EUR
8	Florian Troesch	Design and Optimization of Distributed Multiuser Cooperative Wireless Networks ISBN 978-3-8325-2400-5 42.50 EUR
9	Celal Esli	Novel Low Duty Cycle Schemes: From Ultra Wide Band to Ultra Low Power ISBN 978-3-8325-2484-5 42.00 EUR

10	Stefan Berger	Coherent Cooperative Relaying in Low Mobility Wireless Multiuser Networks ISBN 978-3-8325-2536-1 42.00 EUR
11	Christoph Steiner	Location Fingerprinting for Ultra-Wideband Systems. The Key to Efficient and Robust Localization ISBN 978-3-8325-2567-5 36.00 EUR
12	Jian Zhao	Analysis and Design of Communication Techniques in Spectrally Efficient Wireless Relaying Systems ISBN 978-3-8325-2585-9 39.50 EUR
13	Azadeh Ettefagh	Cooperative WLAN Protocols for Multimedia Communication ISBN 978-3-8325-2774-7 37.50 EUR
14	Georgios Psaltopoulos	Affordable Nonlinear MIMO Systems ISBN 978-3-8325-2824-9 36.50 EUR
15	Jörg Wagner	Distributed Forwarding in Multiuser Multihop Wireless Networks ISBN 978-3-8325-3193-5 36.00 EUR
16	Etienne Auger	Wideband Multi-user Cooperative Networks: Theory and Measurements ISBN 978-3-8325-3209-3 38.50 EUR
17	Heinrich Lücken	Communication and Localization in UWB Sensor Networks. A Synergetic Approach ISBN 978-3-8325-3332-8 36.50 EUR
18	Raphael T. L. Rolny	Future Mobile Communication: From Cooperative Cells to the Post-Cellular Relay Carpet ISBN 978-3-8325-4229-0 42.00 EUR
19	Zemene Walle Mekonnen	Time of Arrival Based Infrastructureless Human Posture Capturing System ISBN 978-3-8325-4429-4 36.00 EUR
20	Eric Slottke	Inductively Coupled Microsensor Networks: Relay Enabled Cooperative Communication and Localization ISBN 978-3-8325-4438-6 38.00 EUR

21	Tim Rüegg	Low Complexity Physical Layer Cooperation Concepts for Mobile Ad Hoc Networks ISBN 978-3-8325-4801-8 36.00 EUR